The new Discus Book

Bernd Degen

I would like to thank the many friends and readers from bede Verlag for the many photos of their discus fish that they have made available to our archive. I have always had many friendly contacts in Asia, especially with the members of the Singapore Discus Club and the Malaysia Discus Club. Nathan Chiang from Fish Magazine Taiwan and I have always had a particularly deep friendship. Helge Muß-topf accompanied me for many years until today as a good friend and discussion partner in discus matters.

Dieter Untergasser has set a milestone with his book, only available in german. Known from many publications by bede Verlag, he has contributed his large disease table to this book.

PAULUS
AQUA SCAPING

Thanks also go to Alexander Paulus for his cooperation on the book.

Explosion
A painting by Lore Degen

1984 was the year when our bede publishing house was given its name by chance. My wife Lore and I had decided to have my first discus book produced. A discus book had to show color pictures of these majestic fish. Despite the high cost of color lithographs at the time, we decided to produce a "real" book. This led to the title "The new Discus Book". There was hardly any competition for this book on the German market, as the discus fish was probably not an interesting business for German publishers. But we wanted to do it right and registered this book title with the Booksellers Association. As a result, we received a phone call: "What is the name of your publishing house?" - a good question that we didn't have an answer to ourselves. My wife responded spontaneously and ingeniously: she simply took my initial letters B and D. "And how do you spell that?" was the counterquestion. "Quite simply - bede!" My wife Lore took over the management and we ended up with a total of 600 book titles on pets of all kinds. There was almost no pet, no matter how exotic, for which there was not one or more bede books. The two discus books were followed by dozens of other discus books and also a special discus magazine "Discus live". After the sale of the pet books to Ulmer Verlag, bede Verlag became Degen Mediahouse - Europe's largest online publisher on the subject of bicycles with Velomotion.de, followed a few years ago by Scout Medien GmbH, which today publishes specialty books from various fields. And now it is my new discus books that are continuing this story. Incidentally, this book is also available in English.

As I hadn't published a discus book for a very long time, it was time, despite the Internet age, to make a "real" book again. Why not two great books? That was the question that moved all of us in the family and so, in addition to this book, we planned a unique, gigantic work about the most interesting discus fish from many years of passionate love for these kings. Friends from all over the world contributed photos and thousands of discus photos were stored in the former bede archive anyway. But it had to be something great and that's why the book "888 Discus" was created. A unique book with 888 exempted discus and about 100 accompanying photos for explanation. There are only 888 numbered and signed copies in the best quality worldwide.

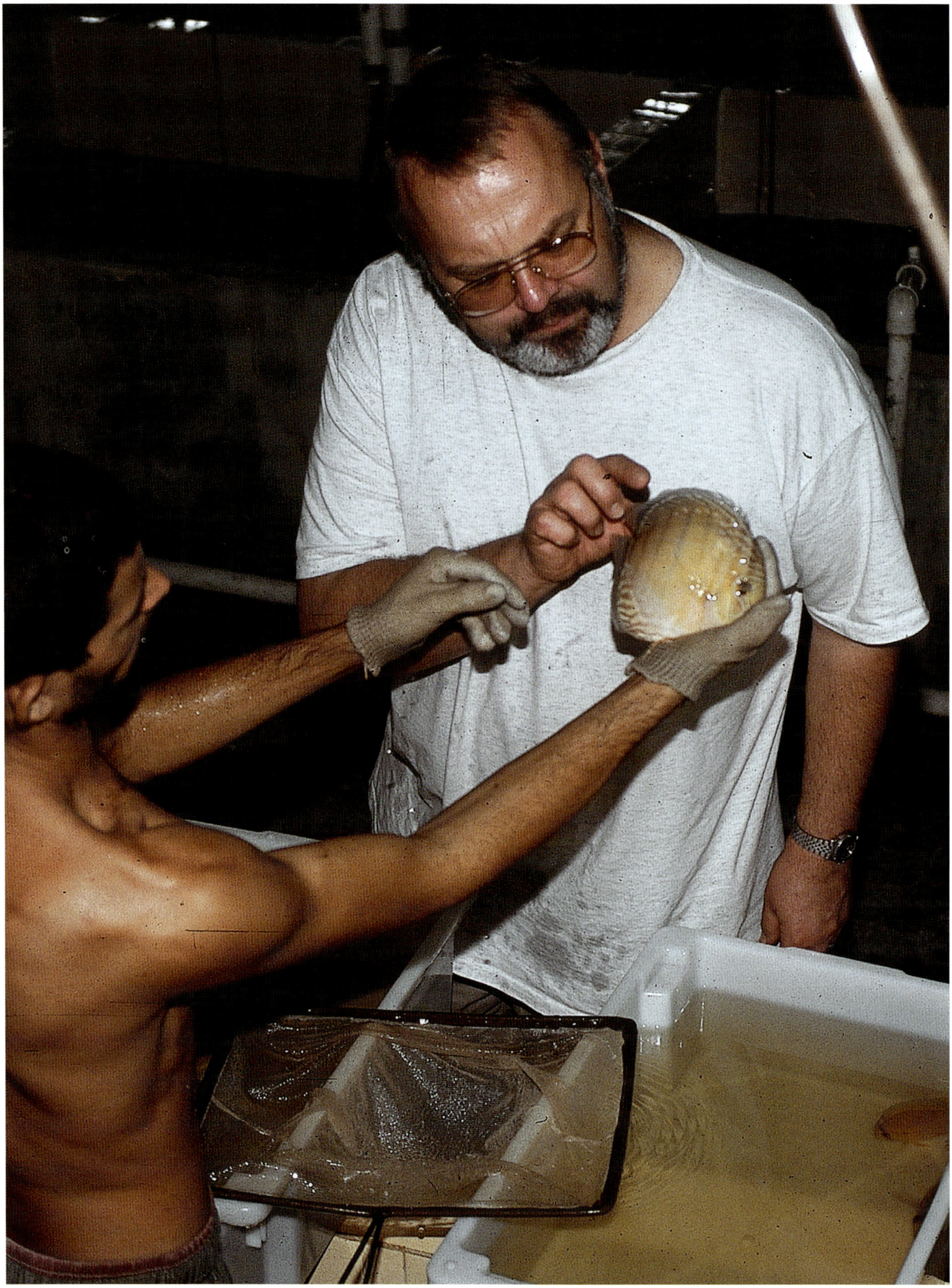

In Brazil I was able to choose many discus for breeding.

It was exactly 50 years ago in 2024 that I started writing about discus fish. In 1974, my first "booklet" about discus fish and other aquarium inhabitants was published in collaboration with TFH Publications in the USA and Vitakraft Germany. Today I am perhaps writing the text for my last discus book? With a little melancholy, you can believe me, because writing about discus fish for 50 years is a damn long time.

I have written hundreds of pages and met hundreds of interesting people in the process. People near and far. I have been able to travel to many countries and search for interesting topics about the discus hobby. It was a great part of my life and I was really lucky to have so many positive experiences. It didn't matter whether I drove to a breeder in the Taunus region with a bulky flash system, set up my system and tripods and shot three slide films, while at the same time dragging the proud breeder's little daughter in front of the aquarium and persuading her to throw food into the fish so that I could take a nice atmospheric picture.

On the other hand, this could have been a flight to southern China during the SARS epidemic, long before Corona, when you had to walk through disinfectant broth directly in the airport and an electronic thermometer was used to measure possible fevers from a distance. It all looked a bit ridiculous. But it really was a dangerous time and I was asked several times why I had to fly here now of all times. Because of the discus fish and great reports for my readers, of course. I have attended dozens of discus champion-

One of my first discus breedings about 50 years ago.

In 2004, my son Michael founded Tropical Germany with Bartek and Tadeusz Ogrodnik.

ships around the world as a judge and it was always important to me to support the organizing clubs financially and not to pay for plane tickets. Friendly cordiality was much more important to me than reimbursement of costs.

Thanks to Norbert Zajac as president, I was able to accompany the world's largest discus championships for many years and show the discus world that Germany was the birthplace of the worldwide love of discus!

I was always a listener during my visits to many discus cellars, because being a teacher is not the best thing to do and book authors in particular have to prick up their ears everywhere to find out something new and interesting. Listening is the magic word.

I kept thinking that everything had already been written about discus fish. But new things are constantly being added. This hobby has changed and will continue to do so in the future, but there will always be something interesting.

I hope that you will enjoy this discus book. Should we ever meet at an exhibition or in a pet store, please contact me, I would be delighted to hear from you.

Have a look at scout-medien.de for the unique 888 Discus book, or write to me on Facebook, that would also make me happy.

Dear discus enthusiast, I wish you continued enjoyment with your hobby and all the best for the future!

Yours, Bernd Degen

There have been and still are great trophies at international competitions.

Amazonia - World of the discus fish

Discus fish inspire hobby aquarists all over the world. The genuine discus fish from Amazonia are very different in color, in contrast to the many captive-bred varieties. Breeders, especially in Southeast Asia, have put many color variants on the market over the last forty years. As a result, the classic wild-caught varieties have been pushed into the background.

In recent years, however, there has been something of a comeback and the run on unique wild-caught fish has started again. There were various color nuances from the different rivers of Amazonia, which were, however, sometimes very small within the areas. In fact, all the discus fish shown in this book are unique representatives of their species. If you look closely, you will see the subtle differences within the color variants. The different fishing areas, which in principle are always rivers and lakes, help to distinguish the discus fish. Numerous trade names have now been established. The magnificent discus fish imported to us from Brazil are the true kings for us aquarists. After their discovery and subsequent import and distribution in the hobby aquarium, keeping them was initially reserved for a few specialists. This was partly due to the very high prices and partly due to the complex care required. It took many years before a few breeders managed to breed successfully. Many years later, it was possible to breed wonderful captive-bred discus from wild-caught strains. The wild-caught discus were increasingly

There are numerous small fishing boats on the rivers. Discus fish can be caught in the tributaries in particular. A brown discus at the top and the typical Heckel Discus below.

Depending on the location, there are different color variants of discus fish in international trade. The names are usually derived from rivers.

replaced by much more colorful offspring. The turquoise discus became fashionable and the demand for wild-caught discus fell dramatically. Great red-turquoise discus were bred in Germany and these replaced the less colorful wild-caught discus. The many new high-breeding colors from Southeast Asia in particular contributed to the fact that wild-caught discus were almost completely forgotten.

However, wild-caught specimens were always important for crossing into color variants, as this was the only way to perfect the sometimes gigantic new breeds. Inbreeding occurred time and again and it was therefore very important to cross wild-caught animals from time to time in order to avoid permanent damage to the offspring. The offspring strains became better and better, the colors stabilized and the typical breeding characteristics were retained. In recent years, there has finally been a turnaround - wild-caught birds have been imported more frequently again.

Although the numbers are small compared to the offspring, interest is growing worldwide.

Many internet publications and discus clubs in social networks have also contributed to this. More and more wild-caught fish with very interesting colors from various river regions are being imported today and in this book we show you how differently the discus fish can be marked in the individual regions. I would like to thank my friend Hudson Crizanto, who has been exporting the most beautiful discus fish from Brazil all over the world for many years. Let yourself be inspired by the differences, but also the common characteristics of the individual representatives of these regions. The demand for beautiful and interesting wild-caught fish is increasing worldwide. Let‘s hope that the home of the discus fish is not destroyed even faster by overexploitation. Responsible breeders are called upon to breed discus fish in their natural form and preserve them as pure species.

Catch & care of wild fish

The capture and export of wild fish is also regulated in Brazil and is subject to state protection. The environmental office IBAMA ensures that permits and controls are in place for fishing and exporting discus fish.

The exporters with larger farms have the best contact with fishermen in the various villages near the river and the fishermen work together with the exporters because they know the best fishing spots.

Hudson Crizanto Gonçalves owns a large farm and exports acclimatized wild discus from various fishing grounds. He is usually involved in the organization and execution himself and sorts the best wild fish.

Catching discus fish is much easier, especially during low water levels. During the rainy season, the water is too high and targeted fishing is hardly possible. This is another reason why the exporters have to keep a stock of fish on the farms, which can then be sold during the rainy season.

Hudson has good contacts with his fishermen and usually goes fishing with them. The discus are encircled along the shore with the help of set nets and then the men begin to catch the discus individually with hand nets in order to inflict no or as little damage as possible to the fish's sensitive mucous membranes. During this process, the gillnet is pulled tighter and tighter. Tree roots or branches often have to be removed, which can make the catch very laborious. Another method is night fishing. As the discus fish usually sleep in the shore area, the men use small canoes and shine a light into the water. As soon as they see the discus standing still in the water, they can catch them more easily with the hand net.

The caught discus are kept at the collection point or in the boat in the typical plastic tubs. If there is a mooring for the fishing boat nearby, the discus can also be packed for onward transportation and taken to the farm.

On the farm there are then appropriate holding tanks, which are often still made of concrete, or the discus are sorted into groups according to catch area or type and prepared in holding aquariums. This often means appropriate quarantine or treatment. Now the discus should also be accustomed to food, as it sometimes takes weeks before they are exported.

In the shallow waters of rivers and lakes, fishermen dive for discus with hand nets.

Catching with a hand net is gentler on the discus.

This just caught discus shows a very intense coloration.

A simple filling station in the widely ramified Amazon region.

Fishing life on the small boats is hard work.

These wooden racks are used to keep the fish until they are transported away.

In addition to gasoline, there are also a few little things to buy.

After the work is done, we have dinner and the next day we take the work away in polystyrene boxes.

The discus are always loaded into cartons from the place where they are caught.

Onward transportation to the large farm and beer for company.

The catchers' life far away from home is monotonous.

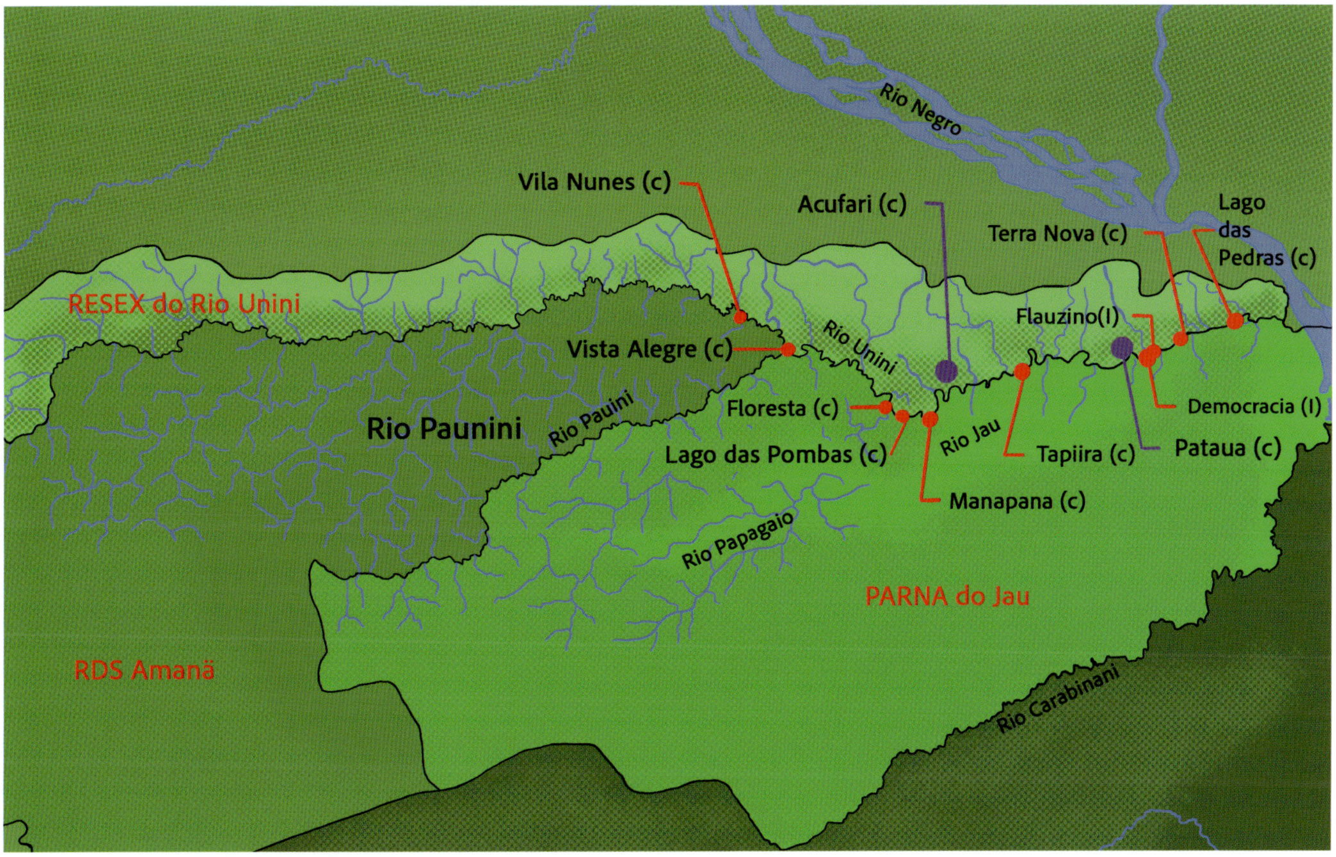

An overview map of where Hudson collected the wild catches shown here. Some settlements in red, special natural areas in purple.

Problems with new imports

Freshly caught discus have to be kept and cared for on the farm for a long time. Only then can these magnificent specimens be shipped.

The situation with wild-caught imports has improved considerably in recent years. This is primarily due to the fact that exporters now obtain a much higher price for discus fish than was the case twenty or more years ago.

Now the exporter takes much better care, packs better and so some of the problems that used to exist are no longer relevant. The brown, green and blue discus are the least complicated - they come from waters with a higher pH value. The Heckel Discus is different. It has a certain special status. The Heckel Discus is the discus that can always and immediately be recognized by its typical Heckel band. Symphysodon Discus was first described as the Heckel Discus in 1840.

To this day, this wild specimen has remained the most special discus and continues to puzzle us discus enthusiasts. It is rarely found in specialist shops and top qualities are very rare anyway. This is also due to its somewhat more difficult husbandry, which is more demanding than is the case with other wild-caught and captive-bred discus. It is and remains a special discus for discus enthusiasts.

A great deal has already been published about the Heckel Discus, and specialists are happy to try and claim that they have successfully bred the Heckel Discus.

However, this endeavor must remain doubtful, as it has never really worked in the long term or only small successes are known. Mating with blue or brown wild-caught fish has been proven in the wild and so-called "Red Marron" Discus have appeared time and again.

These were brown and had a clear stern stripe. Today there are beautiful imported discus that show a strong black central stripe, but are not referred to as Heckel. They can show turquoise stripes of varying intensity, which indicates natural crossbreeding. Wild-caught specimens with individual Heckel stripes also continue to amaze wild-catching enthusiasts. These crosses fetch higher prices.

The unique Heckel Discus is found exclusively in the blackwater rivers of Amazonia.

The middle and lower Rio Negro with its tributaries is the typical habitat. Rio Urubu, Rio Jatapu, Rio Trombetas, Rio Nhamundá and Rio Abacaxis are some of the rivers that can be associated with the origin of Heckel's Discus.

From an aquaristic point of view, these blackwater rivers have extreme water parameters. Various authors and several of our own measurements at different times have repeatedly determined an average value:

For the pH value, values between 4.0 and 5.0 can be classified as average - extreme pH values from an aquaristic point of view. Let's assume average pH values of 7.5 to 8.5 for our tap water. These are huge differences and discus that would be transferred from such acidic water to such high pH values would have considerable mucous membrane problems. problems. In extreme cases, they could even die of shock. In terms of electrical conductivity, the situation is also dramatic from an aquarium point of view. Our tap water is usually hardened and then has conductivities of several hundred or even over a thousand microsiemens. In Amazonian black water, the conductivity is usually below 20 µS/cm! A value that we can hardly comprehend in the aquarium. If we were to desalinate the water like this, the pH value would become so unstable that we would constantly have problems problems adjusting the pH value to some extent.

We therefore need a starting conductivity of at least 50 µS/cm in the aquarium for Heckel wild-caught fish. This conductivity then slowly rises in the aquarium due to food residues and excretions from the discus. We are happy if we can maintain an electrical conductivity of 100 to 200 µS/cm in the water.

At these values, we can get discus fish to breed well. However, there is one exception - the Heckel Discus. It places the highest demands on its water and especially on the breeding water.

Brown discus are once again in great demand, because an intense brown with a red tint brings high sales prices.

Heckel Discus offered for sale here are wild-caught. They have already become accustomed to different water values and may also look quite good.

But somehow, as a buyer, you always have the feeling that something is missing or not quite right. This is also one reason why Heckel Discus have not become big sellers. Unfortunately, it is the case that Heckel wild-caught fish have the highest losses from catch to home aquarium. The other wild-caught fish do not have these losses as they have much lower or more normal water requirements. In the wild, they live in waters with pH values between 5.5 and 7.0 and a conductivity between 20 and 50 µS/cm, i.e. quite acceptable values that can be easily reproduced in the aquarium. This means that the stress between catching, exporting and the home aquarium is significantly lower than with the Heckel Discus. From our own experience with the catchers and at the export station, we can say that Heckel Discus are transferred too quickly from the acidic black water to other water. Then there are already many failures. Freshly imported Heckel Discus always have problems. It is to be hoped that such fish will end up in the hands of experienced enthusiasts who are prepared to set up typical Heckel aquaria.

The main problems are always the mucous membranes, especially the mucous membranes of the eyes. Heckels very quickly suffer from irritation of the mucous membranes or injuries which, in the case of the eyes, lead to a white coating. These infections do not heal under unfavorable water conditions, leading to blindness or even loss of the eye. First of all, the pH value must be slowly reduced. Constant measurements are important. Only if there is no rapid improvement under optimum water conditions should preventive treatment be carried out. Our drug of choice here was always Cotrim forte, whereby we used one capsule per 100 liters of water. The prescription medication remained in the aquarium for three days, followed by a partial water change and good filtration. Contact your vet for the prescription.

So far, all Heckel Discus have been brought back to top form in this way. They thank you with pH values of 5.0 to 5.5 and a conductivity of around 100 µS/cm with the most beautiful colors. Anyone who has ever seen Heckel Discus in top condition will fall for these beauties.

Breeding is a different and more difficult topic. Here, pH values of 5.0 or slightly below and conductivities below 50 µS/cm are required, and the temperature should be between 29 and 31 °C. Give the Heckel Discus a try! Things are much better with the other wild-caught discus. They come from waters with higher pH values and are therefore less problematic. These new imports are usually kept by the importer for a few weeks before they reach

On the left a reddish-brown discus, on the right a Heckel with intense blue coloration in the head area.

the market. By then they have also been largely adapted to our water values. In the absolute luckiest case, there are no problems with the water.

Feeding can be more difficult. Patience and variety in feeding can help here. Offer as many types of food as possible to find out which ones are accepted. It is not always the live mosquito larva that is taken immediately.

The result is usually particularly disappointing here, as the discus are not used to such twitchy food. It is therefore hardly surprising to see freshly imported wild-caught fish from a wholesaler eating spirulina flakes on the second day. The adaptation of captive-bred fish is definitely easier. If these discus are imported and come from South East Asia, they also have a very long transportation time behind them. The exporters, who have been in the business for a long time, know how to pack and ship discus fish, which is why there are hardly any losses nowadays. However, appropriate quarantine is also important here.

Three perfectly marked blue-headed Heckles.

Somewhat more difficult to maintain and therefore a challenge for experts.

Unique beauties from Amazonia

Hudson Crizanto, an old discus friend from Brazil, is the owner of a major aquarium fish export company. Hudson loves his Amazonia. The endless expanse, the diversity of the animal world, but he is particularly dedicated to the careful export of discus fish.

Hudson loves to keep in touch with the fishermen in the villages on the edge of the vast riverbank, with the multitude of rivers and to go fishing with the local fishermen himself. He spends many weeks during the season on the small fishing boats to keep finding new varieties of discus fish and to surprise his customers all over the world with particularly beautiful and interesting discus.

Here is a current selection of beautifully marked discus from the last fishing expedition, which have already spent some time at the farm. Every day of care at the farm is visible, as damage to the mucous membrane or minor injuries heal quickly.

Depending on where they are caught, these discus are given a location designation so that the individual typical differences can be easily recognized. Exporters like to use these names for their sales lists. Some of these names then become internationally recognized names and aquarists can use them to imagine what these discus actually look like. Ultimately, it is always an advantage when making a purchase if you have good contact and trust in the seller. The internet often helps you to order and obtain a specific discus from a dealer.

i

The following wild catches were designated according to the catch locations and thus represent an individual designation.

Red Curuá

Alenquer Rio Inanu

Discus from the Alenquer region and the Curuá river

After just a few days, the discus in the aquariums at the farm near Hudson begin to show intense colors.

The name Alenquer first appeared around 30 years ago. There were many puzzles about spellings and fishing areas. However, this name was intended to indicate strong reddish-brown discus.

When Alenquer Discus show a blue stripe, they are referred to as Royal Alenquer. A wonderful color combination of blue and red-brown!

Green Tefé Coari Japura

These partly greenish flat discus are called Tefé Discus if they are heavily dotted in red. The more dots, the more expensive.

Green discus that have just been caught and show strong stripes are sold as Royal Green. An absolute dream discus - after weeks of acclimatization!

H&K
Ornamental fish
Atenciosamente.
Hudson Crizanto
Sócio-administrador
+55 85 9 9944-1101
@hekhudsoncrizantobrazil
Hudson Crizanto Gonçalves

The Rio Negro is a mighty blackwater river and home to the Heckel Discus.
This shoal of Heckel has just been introduced and still shows a darker fright coloration.

As soon as some time has passed, these Heckel show a stronger color and also red colored eyes. Both are a sign of health.

Heckel Unini and Rio Paunini

The dark, black fifth central stripe is the typical distinguishing feature of the Heckel Discus.

These Blue Face Heckels can be found in the Rio Paunini, a tributary of the Rio Negro. A particularly attractive blue lining.

Heckel Rio Unini and Rio Paunini

Discus from the Nhamundá and Cuminá River

The Rio Paunini flows into the Rio Unini and then both flow into the Rio Negro. These beauties can be found in this black water with pH values below 4.0.

Heckel Nhamundá Blue

Heckel from the Cuminá and Nhamundá Rivers

Heckels of this type are also known as Nhamundá Discus. This refers both to the river of the same name and to a town in north-eastern Amazonia.

Nhamundá Brown

The name Nhamundá appears for differently marked discus, which can cause confusion; the river basin is so large that there are different types of discus in the same area.

With the advent of the turquoise discus, the brown discus fell into oblivion for a long time. For over ten years, strong reddish-brown discus have been in demand again and are fascinating connoisseurs.

Nhamundá River - Royal Discus

Stripped Royal Discus from the Nhamundá River

Discus mate regardless of color variations. This is why there are also striped discus in the Nhamundá. Turquoise-blue stripes all over the body are called Royal Discus.

Discus from the Nhamundá River

Royal Discus and Semi Royal Discus from the Nhamundá River

The discus arrive at the farm in white plastic tubs. A dark startle coloration is normal. They soon lose the dark coloration and show lighter colors.

Nhamundá Red

Nhamundá Royal

Royal Ica red

These red and reddish-brown variants are found in the huge river basin of the Rio Madeira. Discus types with the dark, black fifth central stripe are called Ica variations. Hudson has already found such rare crosses in many places. Both from Alenquer, Nhamundá, Trombetas, Purus and also among the green discus of Tefé, Jurua and Coari. These Ica Discus are highly sought after by enthusiasts.

Madeira Royal Cross

Royal Cross from the Paraconi River from the Madeira Basin

Red-brown discus from the Paraconi River

DISCUS

50 years
Discus passion

Spring 2026
SCOUT-MEDIEN.DE

888

888 released discus
over 1000 photos
Large format illustrated book

ISBN: 978-3-948309-20-6

History - Back in Bangkok

It happened in Bangkok in 1988, when the famous breeder Kitti Phannaitthi brought the first Pigeon Blood Discus onto the market - and I was there. As so often, I met up with Kitti on his hidden farm in the center of Bangkok; it was almost a ritual between the two of us. We always had to have an interpreter with us. Kitti spoke no English and I - despite many visits to Thailand - spoke no real Thai. First he showed me his novelties, and there they were: the Pigeon Blood. They didn't have a proper name yet, were half-grown, showed more sooty grey than color - and also a pair with young on their bodies. At first I was rather disappointed, there was no sign of red coloration. But there was a flash in Kitti's eyes when he caught the first half-grown Pigeon Blood with the net and showed them to me under strong lighting. Yes - there were indeed red spots to be seen, hidden under all the gray mishmash of color nuances. They looked strange, but fascinating at the same time.

Now Kitti wanted to know. We drove - this time without an interpreter - to a breeder friend. It was already past midnight when we arrived at an old farm far outside the gates of Bangkok.

First Pigeon Blood

We were greeted by loud, wild dogs barking - it was an eerie scene. Out of the house stepped an elderly man in shorts and bare-chested. He greeted Kitti in a friendly manner and me, the "farang" (as the Thais call foreigners) with conspicuous respect. Unfortunately, he didn't speak a word of English either, so everything was done with hand signals and glances. Here, in this place, was the home of the first Pigeon Blood - the first offspring of these gray discus that nobody wanted. The breeder's buyer refused to take them away. But Kitti, the clever fox and great breeder, immediately recognized that there was something special about these fish. Something strange - and perhaps something big.

He took the first fry and the parent pair with him and formed the original strain of one of the most important breeding lines in discus history - a true mutation. These Pigeon Blood became the prototype for a large number of later crossbreeds and had a lasting influence on the world of discus fish. Kitti and his Pigeon Blood changed discus breeding - until today.

Modern Pigeon Blood

Kitti was always a master at breeding animals, so he also had a great passion for dogs, breeding Bernese Mountain Dogs, albino tortoises and albino stingrays, which he was also able to sell for horrendous sums of money.

Buy young fish or larger discus?

Buying discus is a matter of trust, no matter where you buy your discus fish.

When you have decided for the first time that you want to keep discus fish, you will automatically be faced with the question of whether you should buy smaller juvenile fish or larger or even adult discus fish. The answer to this question naturally depends very much on your needs. If, for example, you want to build up a breeding stock and one day breed several pairs of discus, then it is advisable to buy a larger number of good half-grown discus with a body size of at least ten centimetres. Then, after six to eight months of good care, you will have a fine selection of discus fish that can already form pairs and want to start breeding.

It is certainly an advantage if you buy fish from different breeding lines. However, it is good if all new discus fish are purchased within a short period of time if possible, so that a joint quarantine is possible. However, it should be noted that the adult discus fish should also match in color, as it makes little sense from a breeding point of view to mate a blue discus with a red discus later on. If you have the space for several larger aquariums, you can confidently buy two or three schools of 10 to 20 fry each at the same time and then rear them.

Raising discus fish is not as easy as it may seem at first, as setbacks or illnesses very quickly have a negative effect on the body shape of the discus fish. Young fish that have become ill during the growth phase tend to lag behind in body size or become somewhat elongated and therefore unattractive in shape.

If, on the other hand, you want to have a beautiful living room aquarium with easy-care discus fish as quickly as possible, we recommend buying larger discus fish.

This young discus already shows growth defects. The finning and the eyes, which are too large in relation to the body, clearly indicate this. Left: perfectly reared half-grown discus. It pays to have waited a little, even if the purchase price was higher.

Buying the right discus from a large shoal of half-grown flat turquoises should not be a problem. Place them in front, observe and select them bit by bit.

A minimum size of ten centimetres and an age of around eight months or older is recommended. A group of six, eight or ten of these semi-strong discus fish in a suitably large aquarium is a lot of fun and it only takes a few months of good care and feeding for these discus fish to grow into truly magnificent discus. If breeding happens by chance, there is nothing to stop you trying it out and transferring the pair from the community aquarium to a breeding aquarium.

Large and fully grown discus fish of a respectable size are of course very impressive for any observer. As such adult discus can be quickly assessed in terms of their qualities, you really do recognize immediately what you are buying. On the other hand, such adult discus are also correspondingly expensive and then you are usually limited to a few specimens.

Raising this group will certainly produce one or more pairs. Up to this stage, about 50 liters of water per fry is the ideal minimum size for an aquarium.

A beautiful group of almost fully grown red Pigeon Blood. This aquarium has a water capacity of around 1,200 liters, which is already attractive due to its size. The larger the aquarium, the less stress and work for the owner.

Relying on acquiring one or two pairs is always critical and should not be viewed too euphorically, as sex differences cannot be recognized with certainty.

When breeding a group of discus fish of around ten centimetres in body size, it is a good idea to plan at least 50 liters of aquarium water per discus. When buying adult discus, which quickly reach a body size of fifteen centimetres or more, 80 liters of aquarium water per discus is ideal. An aquarium set up two meters long and 60 cm deep and high with eight to ten large discus is a real show. If you then add a suitably large filter tank, there will certainly be few problems with the perfect discus water. Also make sure you have the right lighting for your new discus aquarium. Neon tubes are now obsolete, as they need to be replaced regularly. Today, LED lights are the best option, with Daytime lights providing the greatest comfort and longevity.

Ratio eye - body size

You should apply strict standards when buying discus. Regardless of the size of discus you decide on when buying for the first time, it is always important to make sure that your chosen fish make a healthy impression. Healthy discus are alert and active. They have a bright, vibrant coloration and are by no means shy and dark in the corner of the aquarium.

Of course, they should also be free of external injuries, and then as a buyer you should look at the ratio of eye size to body size. With a little experience, you can quickly determine whether a discus is healthy. Look deep into the eyes of your future discus, in the truest sense of the word. Its eyes must be bright, clear and without defects. Very dark, even almost black eyes indicate illness. Under no circumstances should the eyes be milky, as this is a sign of infection. Today, there are also many albino discus and the quality of the eyes is even more important, as only the beautiful red pupil ensures that these albino discus look good. The size of the eyes in relation to the body size always plays an important role for the expert in recognizing a healthy and perfect discus fish. It is certainly not so easy at first to recognize whether the eye is actually too big for this fish or whether it fits. But if you look at the fish as a whole, you will get a feeling for whether the eye appears too large.

If it really is the case that the discus has an eye that is too large, then this either indicates that the fish is very old in the case of adult fish, or generally that it has survived illnesses that have led to a reduction in body size. Beautiful small button eyes are certainly the ideal case. If smaller or half-grown discus fish show disproportions in eye size to body size, this indicates growth damage that can generally no longer be corrected. Such discus fish will certainly not grow into magnificent discus.

The eye color of discus fish is not uniform. There are discus fish with light yellow, yellow, amber or bright red eyes. A breeding goal is usually to have bright red eyes with a rich, black inner pupil. Almost all shades are possible. Especially in new breeds, there are very light, sometimes almost colorless eyes that are not particularly attractive. There are also color gradations in the eyes of wild-caught fish, for example most Heckel Discus have amber-colored eyes, while green discus often have bright red eyes. Of course, it is always a matter of taste which eye color is preferred. But red is certainly the most popular eye color because it is the most contrasting.

This albino discus has a red coloration of the iris, i.e. the outer edge of the eye, and a strong red coloration of the inner pupil. This makes albino discus easy to recognize.

Gold-colored albino with a very round and even discus shape. Here, however, a very bright iris is visible.

Green tomboy that is younger but has growth defects. Body shape too elongated and eye too large for the body size. The color is also faded.

An older Tefé wild caught that is healthy, but due to its age has an eye that is too large in relation to its body. Discus fish can live for ten years or more in the aquarium.

Discus fish feel uncomfortable with angelfish: pale coloration, too elongated, slower to feed - this affects their body shape and behavior.

German forefather in the 60s

A bulky discus of the old Red Scribbled color form, which I was the first breeder in Taiwan to bring onto the market. Here is an anecdote about our esteemed Dr. Eduard Schmidt-Focke, with whom I had the pleasure of preparing his only discus book in his house and then bringing it to market in German, English and Japanese with bede Verlag and TFH Publications.

The word bulky, when used in connection with discus fish, was a red rag to him and had to be avoided at all costs!

A discus should be circular and also have a circular eye, as here. The finning matches the body shape and size perfectly. A strong head section and the markings should also delight discus fans. Would certainly be a good buy!

Iris strong red and pupil black, but nevertheless a somewhat too pointed head shape. The forehead slopes too steeply from top to bottom.

Here is a good example of a very nice rounded head shape. The eye is also perfect in size.

Holes in the head section and damage

Really deep holes in the head of a discus fish are always a warning sign for any discus lover.

These holes can be caused by an infestation of intestinal parasites, but can also have completely different causes. If they are fairly evenly shaped, smaller holes in the head area, the dreaded hole disease can be suspected. However, if there are more uneven holes that do not appear too deep, then it is often a case of deficiency symptoms. These holes mainly occur when the water quality is not right, the aquarium is overstocked and there is a mineral deficiency in the aquarium water. However, an exact diagnosis is not easy. In any case, such holes in the head section, if they are clearly visible, are a real obstacle to purchase. Very small, fine holes such as pinholes on the head area, on the other hand, are completely harmless and should not be associated with pitting disease. There are also two identical small nostrils in our discus fish.

Damage or minor injuries occur mainly when discus fish are caught and during transportation. However, these minor injuries are harmless and heal within a very short time without leaving any residue. Larger injuries, on the other hand, must be treated quickly, as otherwise the wound may become infected with fungus and heal badly or even leave a clearly visible scar after healing.

Larger injuries to the fin margin, for example, leave permanent visual damage. In the case of discus catches in particular, there are many fish that have been bitten by a predatory fish in their youth and this is always immediately visible due to irregularities in the fin seam. Although such visual defects do not affect the offspring, they do affect the beauty of the fish.

Watch out for a sunken head area

One of the main problems in caring for discus fish is emaciation, which also results in a completely sunken head. Discus fish that start to lose weight are very ill and difficult to get well again. To prevent this from happening in the first place, good water care and a balanced diet are important.

A severe flagellate infestation is usually accompanied by emaciation of the fish.

Flagellates

They are usually present in the fish anyway, but are not really a problem with optimum care. A moderate infestation of flagellates is well tolerated by discus fish.

However, if there is insufficient nutrition or severe stress, the flagellates usually multiply explosively and then damage the fish. Protoopalina is a very large intestinal flagellate that only occurs in discus fish. These intestinal flagellates are preferably treated with the prescription drug metronidazole. You will need to consult a vet and describe the symptoms. However, there are now also very effective products available from pet shops. Ask your pet shop about this, as there is always something happening in this specialist field.

When using metronidazole, a dosage of one gram per 100 liters of water is used for three days and no longer. The temperature can then be raised to 33 °C. This dosage is very high, but has proven to be effective in practice. After treatment, a large partial water change and filtering with activated carbon is advisable. Intestinal flagellates are really difficult to combat and often all that remains for the discus owner is frustration. It is therefore advisable to ensure that the husbandry and feeding are optimized in advance.

One-sided feeding with a beef heart mixture can often become a problem, especially when it comes to feeding. Of course, the fish grow strongly at first, but there is a lack of fiber and carbohydrates. This leads to digestive disorders and intestinal sluggishness. Ideal conditions for flagellates. The intestinal wall is constantly irritated by the food slurry and can no longer rest. If, on the other hand, the discus is optimally digested, intestinal flagellates can only multiply with difficulty.

Massive holes can already be seen above the eyes. These typical holes can also be seen in the area of the gill covers and the mouth. The discus still appears active and nourished, but its condition will deteriorate over time. Other discus in the same aquarium will also be affected.

In addition to the typical holes in the head area, white deposits can be seen on the eyes of this discus, indicating a bacterial infection. The discus is also clearly emaciated. Only persistent treatment can bring success here.

This discus is already very old, but has no head holes, so it is healthy, even if it appears somewhat emaciated. Just a sign of old age. If a discus has lived in the aquarium for many years, perhaps ten or more, it will naturally lose a little weight but still remain lively.

For comparison, a well-fed Blue Diamond of prime discus age. Strong head section, and the hole is a typical nostril - not to be confused with hole disease! Unnecessary treatment for the disease would be stressful. So please observe carefully before adding medicine to the aquarium.

Emaciated discus -
feces, a distinguishing feature

You can usually tell what a discus has eaten by looking at its droppings. For the experienced observer, the droppings that discus fish deposit in the aquarium are a typical indicator of whether the discus is healthy. Dark excrement that falls off the discus without any problems and sinks to the bottom is an indication of health. If a lot of colored food is fed, it is completely normal for the droppings to take on a reddish color. However, it is important to note that in healthy discus fish the droppings always have a firm consistency and fall quickly from the fish to the bottom in pieces. However, if the excrement remains attached to the fish for a long time and does not fall off immediately even with more vigorous swimming movements, problems must be expected.If the fish drag white, gelatinous excrement in long threads behind them, often for several minutes, as they swim through the aquarium, this is a sign of digestive problems caused by flagellate infestation. This white feces is a real sign that there is most likely an infestation of intestinal parasites. As these parasites are difficult to identify and are also difficult and time-consuming to combat, we strongly advise against buying such discus fish. So take a look at the color or consistency of the droppings floating around in the aquarium. If you see white threads of excrement, this is an absolute warning sign.

This discus already shows some unfavorable defects. The caudal fin is already affected and has partially fallen off. The eye is cloudy and the iris has a very dark coloration. Antibacterial treatment is recommended here.

◂ *On the left you can clearly see the black feces of this healthy discus.*

Green wild-caught fish after import. The middle discus shows some coating on the eyes and should be treated. Quarantine and close observation of the discus is particularly important for freshly imported wild-caught fish in order to recognize problems quickly. Such a coating on the eyes can be treated with antibacterial agents.

White excrement, which is very viscous and gelatinous, indicates a flagellate infestation and must be treated. Although the discus will still eat, they will lose weight. Smaller discus in particular suffer very quickly and waste away. Adult discus can survive for a long time with such flagellates. The droppings of healthy discus must be dark and fall off the fish quickly. Under no circumstances should droppings be dragged behind the discus for any length of time.

In Amazonian rivers, our discus fish naturally also get sick and are plagued by parasites. The big difference, however, is that in an aquarium with relatively little water, a completely different disease pressure can prevail in comparison. Optimal water care is therefore very important. An old saying goes that the larger an aquarium is, the less work it is and the fewer problems there are. Take a look at these Heckel wild-caught fish that are swimming in a well-established aquarium, but don‘t seem to be feeling quite right yet. What could be the reason for them being too dark and with too little color in the aquarium? The many roots provide plenty of shelter, but also constantly release a lot of substances into the water. Wood tends to decompose slowly and then the water quickly smells musty. It would certainly be better to use less real root wood. Nowadays there is beautiful artificial wood that can no longer release any substances into the water. The pH value of the water may also be too high for Heckel Discus. These wild-caught fish live in waters with an extremely low pH value of 4.0 on average. It is difficult to set this low in the aquarium, but a value between 5.0 and 6.0 is feasible and is well accepted.

No.	Appearance	Possible cause	Description/drug/active ingredient
01	Skin abrasion	Catching, rough net	Water conditioner
02	Injury	When fish get frightened or rub themselves	Acriflavine
03	Melting of fins, holes in the head area, hole disease	Mineral deficiency due to mostly deionized water	Special mineral salt
04	Dark coloration of the discus	Heavy metal ions	Water purifier, activated carbon filtration
05	Fin rot, fraying of the fins	Bacteria, water germs	Acriflavine
06	Small, white eruptions on the skin, like pimples	Scaling pocket inflammation	Nitrofurantoin or Cotrimstada
07	Mucosal detachment	Discus search	- Nitrofurantoin or Cotrimstada - Neomycin Sulfate
08	Cotton wool-like threads on the skin	Fungus	Acriflavine
09	Mucous deposits on the skin	Different external parasites	- Common salt (Sodium chloride) - Malachite green oxalate
10	Blue-grey skin coating	Costia	Malachite green oxalate
11	0.1 mm small dots on the skin	Oodinium (velvet disease)	Copper sulphate
12	Up to millimeter-sized and white dots on the skin	Ichthyophthirius multifilii	- Malachite green oxalate - Common salt (Sodium chloride)
13	Demarcated white mucosal thickenings of 5 to 10 mm in size	Chilodonella	Malachite green oxalate
14	Stringy, strip-like thickening of the mucous membrane	Tetrahymena	Malachite green oxalate
15	Rapid breathing	Nitrite poisoning Oxygen deficiency	Sera Nitrit-minus Aerate the water or add oxygen supplement. Find out the cause and eliminate it.

Dosage	Duration of treatment	Remarks
acc. manufacturer		
200 mg/100 l	3 days	Remedies from pet shops are just as effective.
acc. manufacturer	Until the symptoms disappear.	Adjust via the conductance; see text
acc. manufacturer	7 days	
200 mg/100 l	7 days	Remedies from pet shops are just as effective.
50 mg/100 l	7 days	Prescription only
- 40 mg/100 l - 3 g/100 l	- 7 days, min. 50%, water change and then 10 days post-treatment - 7 days, water change	Prescription only Prescription only
200 mg/100 l	5 days	Remedies from pet shops that also contain copper are more effective.
- 1 g per 10 l / 1 x on the first day - 5 mg/100 l, every 2nd day the half the dose every second day	5 days	Remedies from pet shops are just as effective.
5 mg/100 l, add half the dose every 2nd day	3 days	Remedies from pet shops are just as effective.
Concentration in the water with an accurate copper test to 0.3 mg/l, daily measure and supplement daily.	5 days	Remedies from pet shops are just as effective and are toxic to sensitive fish and lower animals.
- 5 mg/100 l, add half the dose every 2nd day - 1 g per 10 l / 1 x on the first day	7 days 7 days	Remedies from pet shops are just as effective.
5 mg/100 l, add half the dose every 2nd day	5 days	Remedies from pet shops are just as effective.
5 mg/100 l, add half the dose every 2nd day	5 days	Remedies from pet shops are just as effective.
10 ml/20 l acc. manufacturer	Can be re-dosed daily.	Pet shop

No.	Appearance	Possible cause	Description/drug/active ingredient
16	Chafing at the gills, rapid breathing, gill cover on one side	Gill worms	- Praziquantel - Tremazole - Formalin 35 to 40%
17	Bleeding on the skin	Mostly bacterial infection	- Chlortetracycline - Nitrofurantoin or Cotrimstada
18	Abdominal distension Abdominal distension, googly eyes, protruding scales	Dropsy Dropsy	- Nitrofurantoin or Cotrimstada - Chlortetracycline
19	Milky white, slimy feces	Flagellate infestation	- Metronidazole - Nitrofurantoin or Cotrimstada
20	Slimy feces, refusal to feed	Worm infestation, nematodes, Capillaria	Levamisole, drug: Citarin®-L10% or Flubenol® 5%
21	Slowed growth	Oxyuris, discus pinworm	Flubenol® 5%
22	Slimy white excretions	Tapeworm infestation	- Praziquantel - Tremazole
23	Emaciation, knife back	Bacterial infestation or parasite infestation of the intestine, inadequate nutrition	Diagnosis and then carry out the appropriate treatment.
24	Eye opacity	Bacterial infection Worm infestation, fungal infection	Nitrofurantoin or Cotrimstada No treatment possible
25	Header	Swim bladder inflammation	- Cotrimstada® Forte - Nitrofurantoin or Cotrimstada

Dosage	Duration of treatment	Remarks
- 500 mg/100l - according to manufacturer - 8 ml/100 l over 10 hours in a separate container	Add again on the 3rd and 5th day.	Pharmacy-only, 50 % water change before each new dosage. The fish to be treated must not have the slightest skin lesion.
- 3 g/100 l in a separate container container - 50 mg/100 l	3 days	Prescription only Prescription only
- 40 mg/100 l - 3 g/100 l in a separate container	Treat for 7 days, add more when changing the water 3 days	Prescription only Prescription only
- 1 g/100 l - 50 mg/100 l	6 days with temperature increase to max. 32 degrees Celsius	Prescription only
200 mg/100 l	3 days, 50% water change, 8 days break, 3 days post-treatment; 21 days, 3 times at intervals of a week	Prescription or Nematol from the pet shop
200 mg/100 l	21 days, 3 times at intervals of one week	Prescription only
- 300 mg/100 l - according to manufacturer	3 days, 10 days break, again 3 days	Only available in pharmacies. When using liquid preparations, the water often becomes very cloudy. After 12 hours, 80% of the water must be changed.
		Feed a varied diet with high-quality food.
50 mg/100 l	7 days	Prescription only
- 1 tablet/40 l in one separate container - 50 mg/100 l	7 days 7 days	Prescription only Prescription only

Chafing and rapid breathing

If discus fish feel unwell or even have skin parasites, they start to rub themselves. This then looks as if they are trying to rub these annoying skin parasites away on furnishings. Usually nothing can be detected, only the strange behavior of rubbing against furnishings is noticeable. Such discus fish must always be quarantined after purchase and then treated with an appropriate medication against skin parasites.

Healthy discus fish should breathe calmly and evenly. If you are standing in front of an aquarium with 100 young fish, each four centimetres in size, and they are all breathing rapidly, you don't need to panic straight away, as this rapid, intermittent breathing is often only due to heavy feeding. Take a look at the bellies of these small discus fish. Are they full? Well, then it's clear that these small, voracious discus fish simply have to breathe so quickly because they have filled their stomachs to bursting point. So there's no need to worry and no need to be afraid of any gill worms, because this rapid breathing after intensive feeding is completely normal, especially in younger discus fish.

However, the situation becomes much more difficult if it is noticeable that many discus fish in an aquarium only move one gill cover and leave the other pressed tightly against their body. This one-sided breathing indicates a gill problem. It is usually an infestation of gill worms that makes itself felt in this way. However, it is only possible to say more precisely after a gill smear and examination under the microscope.

Gill worms are still annoying parasites of discus fish and they are also very difficult and time-consuming to combat successfully. There are many good preparations available from pet shops for most fish diseases. Ask for advice. If there is no heavy feeding and the discus fish are still breathing very quickly and heavily, this may indicate that the aquarium is overstocked and there is an oxygen deficit in the water. Larger discus fish should always stand in the aquarium breathing calmly and moving both gill covers evenly. In healthy adult discus fish, the normal breathing rate is around 60 gill movements per minute. In juvenile fish, the normal range is up to 90 gill movements per minute. If a significantly higher respiratory rate occurs and this is not due to excessive feeding, it may also be due to a too low or too high respiratory rate. the pH value of the water is too low or too high, or the oxygen content is really too low. You need to check these two water parameters.

Discus fish also tend to breathe more heavily if the water temperature is too high. If, for example, the fish are treated for diseases at a higher temperature, the water must be aerated better, otherwise a critical breathing rate can quickly occur.

Gill and stripe defects

Whole shoals of discus damselfish often have more pronounced gill cover defects. It can happen that the gill covers are shortened or that the gill cover is bent upwards. These gill defects are not only unsightly, they also rule these discus fish out for further breeding. It is not necessarily the case that the parent fish already showed this defect. Entire broods have unfortunately had to be killed off because they had gill covers that were too short, reduced in size, stunted or bent. In very small discus fish, this deficiency is particularly noticeable if there is poor aeration in the aquarium and a lack of oxygen in the water. If the aquarium is overstocked and there is a high feeding rate, the water can quickly become overloaded and such defects can quickly occur. Bacterial explosions in polluted water quickly lead to a consumption of oxygen and then to a reduced blood flow to the gills, which results in such deformities that cannot disappear later in growth. Such damaged discus fish are actually doomed. Isolated, even one-sided gill defects can occur from time to time and if you discover such a fish, you should not buy it or use it for further breeding.

Stripe defects in discus fish are an age-old issue and to this day there is no conclusive solution as to how such stripe defects can occur. Such stripe defects even occur in genuine wild-caught fish. Normal discus fish have nine dark vertical stripes. Depending on the basic colouration and condition of the fish, these dark vertical stripes can be strong, weak or not visible at all. However, there are also color variants of new breeds from Southeast Asia that have thirteen, fourteen or even fifteen vertical stripes. On the other hand, there are also breeding forms such as the Blue Diamond Discus, in which the vertical stripes have been completely bred out. These fish no longer show the dark stripes at all.

Ideally, the vertical stripes of the discus fish should be evenly vertical. The distances between the individual stripes are the same, so that a regular pattern can be recognized. Unfortunately, errors in the vertical stripes are becoming increasingly common. This means that one or more stripes are interrupted and sometimes run diagonally into other stripes. This results in an uneven pattern, which is particularly noticeable when the discus fish show the dark stripes more strongly. There have been discussions for years about the reasons why discus fish show such stripe defects. Claims or assumptions that stripe defects are caused by strong pH changes during water changes or unfavorable water conditions in breeding aquariums must remain speculation, as stripe defects are also repeatedly observed in wild discus catches. So if these stripe defects already exist in nature, the formation of stripe defects cannot be caused exclusively by keeping them in the aquarium.

Nevertheless, it is conceivable that changes in the water balance of large rivers, flooding during the rainy season and the resulting changes in pH value are also the cause.

These five breeding discus show what aberrant stripe patterns are possible. By nature, discus have nine dark vertical stripes, which are either not visible, faintly visible or strongly visible, depending on your mood.

Classic wild discus with nine vertical stripes.

Typical Heckel with reinforced first, fifth and ninth stripes.

Offspring Snake with 15 vertical stripes. The facial pattern is typical.

Intensely colored Blue Diamond in which the stripes have been bred away.

Extremely extended fins

Today we are actually used to it and probably consider it quite normal for our discus offspring to have fairly good finning when fully grown. By good finning we mean that the fin fringes are very large and give the discus an impressive appearance. If we make a fair comparison between adult captive-bred discus and adult wild-caught discus, we should notice that the wild-caught discus have significantly shorter or stockier fins. The caudal fin in particular is smaller and not widely fanned out, but runs almost straight and parallel. The captive-bred discus, on the other hand, have very magnificent fins.

So far so good. This could be the result of keeping them in the aquarium and the result of a completely different diet, or it could be the result of genetic predispositions that have been consolidated over time in our captive-bred discus through breeding. These are all very plausible explanations and nothing to worry about. For this reason, it is usually very easy for the expert to distinguish a wild-caught discus from a wild-caught captive-bred one almost one hundred percent. This is because the wild-caught offspring, which have been swimming in our home aquariums for a year, already show clear differences in their finning compared to the parent animals that have grown up in Amazonian rivers. One of the main causes is probably feeding.

In the wild, discus fish usually live at subsistence level as far as food is concerned.

There is not always an abundance of food and, due to the different water levels, the food supply also varies greatly. During the dry season, this can lead to longer periods of starvation. During the high water period, things are better again and the fish get a good supply of seeds or small animals. That‘s just the way it is in nature. Naturally reared fish always grow more slowly, develop more slowly and are generally somewhat smaller and slimmer than their counterparts in the discus aquarium that have grown up under human care. Let‘s take a look at the trout that lives in a stream, where it feeds on approaching insects, and the trout that swims in a fattening farm and is fed several times a day exactly according to a feeding plan. There must be differences, especially if high-performance feed is fed on a fattening farm.

We also feed high-performance food in our aquariums. Don‘t we also swear by the best feed mixtures of all time and offer them to our pets? Don‘t we go to the specialist stores and look for the biggest mosquito larvae to whet the appetite of our discus fish? Of course we do and we are all too happy to sit in front of our aquarium and watch our discus fish fill their bellies. If the discus is doing well, humans are happy. Wherever there is an abundance of food, the animals also have corresponding weight problems. Nowadays, dogs and cats have to see a specialist for nutritional issues and there are even some countries, such as Japan and the USA, where there are special gyms for dogs.

Fortunately, there are no gyms for our discus fish yet, but we shouldn‘t be surprised if our fish are doing so well that the heavy feeding also has an effect on their appearance. The discus fish grow up faster than in the wild, spawn earlier and, of course, have more developed fin fringes.

While it takes around two years in the Amazon for a discus to become fully grown and sexually mature, it can start spawning in our home aquariums after around ten months with optimum care.

This enormous time difference alone must give us pause for thought. The care in the aquarium therefore has a strong effect on the physical condition of the discus fish.

Discus fish with elongated dorsal fins have been appearing in our aquariums for many years. It is particularly noticeable that the dorsal fin is almost flag-like and pointed. These flag-like, elongated dorsal fins are not found in wild dis-cus catches. It can therefore be assumed that this type of fin shape does not exist in nature. The question now arises as to why these pointed fins are becoming increasingly common in the home aquarium. Even in wild-caught fish, the dorsal fins begin to change somewhat after a long period of aquarium keeping. In Asian dis-cus fish in

A beautiful Pigeon Blood pair that has already raised young. How can you be sure of the sexes here? Size and finning are not reliable characteristics. As a breeder, you can easily get it wrong. There are small males and large females, anything is possible!

Extremely strong exhibition discus, whose pointed dorsal fin is rated positively. However, it is uncertain to draw conclusions about the sex.

particular, this flag-like dorsal fin is more or less pronounced. However, one should not assume that this fin shape is a breeding variant.

There have been breeding attempts to change the fins of many other fish.

Just think of guppies, where we have an incredible number of different fin shapes. These changes are intentional and are achieved through breeding selection. In discus, however, these changes, especially in the dorsal fin, probably have other reasons. In many cichlid species, the males tend to have more beautifully extended fin ends. However, these are sexual characteristics that do not occur in discus. It seems very likely that the fin changes are primarily caused by feeding.

Water influences certainly also play a role. Studies have now even been carried out on humans, with the result that not inconsiderable quantities of hormones have been found in our tap water, caused for example by the considerable increase in the use of hormone preparations such as the contraceptive pill and medicines in recent years. Some of these hormones remain in the water cycle and are reabsorbed by us. In the case of discus fish, however, behavioral symptoms certainly also play a role, as the dominance of males is also largely dependent on hormone levels. If the dominance of the males does not have to be so intense in the wild in the Amazon, because the territory and the natural competitors and enemies are much smaller than in a fully stocked aquarium, this also results in increased or less strong hormone production.

However, this hormone production then has an effect on the design of the characteristics of male animals. One could therefore conclude that if a discus encounters few male competitors in the Amazon, it does not need to optimize its appearance so much for the physical characteristics that indicate maleness. In an aquarium where there are several females and of course other males in a very small space, it is important to show the male sexual characteristics more strongly. This could therefore be a major reason why the males develop their finning more strongly and also show their color more intensively.

It is therefore safe to say that feeding and ultimately the targeted or unintentional use of hormones in the aquarium play a role.

Of course, there is also hormone-loaded food and this food can have an effect. For example, if we feed a lot of beef heart, which is heavily loaded with hormones, this can lead to physical changes in the fish. In South East Asia in particular, a lot of hormones are used when keeping and feeding discus fish. On the one hand, the aim is to make discus fish grow faster and show their color faster, but on the other hand, they also undergo a sex change.

In this pair of discus, the male in front has very elongated dorsal, caudal, and anal fins. Popular with many breeders, but far removed from the natural form!

A young pair from a variant with a high proportion of Pigeon Blood. Sexual characteristics are not present. The discus in the foreground has a very round and pronounced forehead shape, which is often attributed to males.

Interesting coloration, but flaws in the dorsal fin and a small but disturbing nose hump. These features hinder the overall impression in practice.

Young discus with very strong finning. The forehead is very steep and should still grow together so that the shape appears more harmonious. Eye slightly too large!

Strongly pronounced fin fringe, which nevertheless fits well with the circular body shape. The "nose kink" is clearly disturbing and could be partly inherited.

Today, sex reassignment is probably no longer carried out as often, but forty years ago in Hong Kong, for example, it was quite possible to keep discus fish in water with testosterone for a long time in order to ensure that all females were genetically transformed into males.

This sex change is also carried out in commercial fish to achieve faster growth. Imagine what it is like to know that males grow faster and perhaps also show better color. Then it is only natural to try to breed only males. By using testosterone in the aquarium water, it was possible to turn all the discus fish in South East Asia into males. Now people will wonder why this has happened.

The answer is actually quite simple and logical. You have to imagine that when a new discus color was discovered, such as the Blue Diamond Discus, the sex change was a real sensation. Suddenly there were completely blue coated discus fish without any disturbing stripes. Such fully chromed discus fish fetched astronomically high prices in the early days. For the breeders who were the first to offer such discus fish, it was extremely important that it took as long as possible for competitors to breed the same fish. If the market had been flooded with larger quantities of such discus fish, prices would logically have started to crumble.

If it was now possible to prevent the rapid further breeding of the young fish by sexing them, money could be earned for longer. So if the competition got hold of 50 of these fry in order to raise them quickly and then start breeding them, the competition also had to realize that they might suddenly have 50 male discus fish swimming in their aquarium.

It was then impossible to achieve breeding success and the discoverer of this discus color had saved the exclusivity of his breeding color for a longer period of time.

As you can see from these explanations, it is not so easy to judge why discus fish show more pronounced fin extensions than is usual in nature. There are some explanations, but very little evidence. Science may one day be able to help here, but until then we discus enthusiasts will have to decide for ourselves whether we like the discus on offer in this form and want to own it.

Attempts at veiltail or veilfin discus would certainly be less attractive and should be rejected. But this danger is not in sight at the moment and we hope that it stays that way for a long time to come.

This albino discus pair shows no typical sexual differences. It is usually assumed that males have stronger fins in discus of the same size. The shape of the head also does not allow any reliable conclusions to be drawn. Only when the pair spawns can the sex be recognized with certainty.

Why a show aquarium?

Because a discreetly planted and furnished aquarium simply looks nicer!

A group of half-grown discus is a good starting stock for a display aquarium. The first discus should not be much smaller.

Discus fish are almost always only kept in cellars, which is a shame. Why not give it a try with a beautiful show aquarium?

The typical glass discus aquariums without a substrate and without furnishings are certainly justified when it comes to successful breeding, breeding selection, quarantine or simple keeping. However, one argument in favor of a display aquarium is that it simply enhances a living space and makes it more interesting. It is a fascinating sight to walk into a room with an aquarium in it. You will immediately look inside and start to rave about it. The desire to own such an aquarium quickly arises. However, the thought of doing so is usually thwarted by the idea that an aquarium is difficult to look after and is a lot of work. Should there actually be a lot more problems with our discus aquariums? After all, discus fish are among the more demanding fish to keep. However, you will soon find out for yourself that this is not the case if you follow the most important rules.

If you own discus fish, you may one day decide to set up at least one display aquarium in your living room. The cellar may then be home to the keeping system and the most beautiful discus fish swimming in the living room. If you follow a few tips, it is not difficult to take this step.

Here, the swimming area of the discus fish has been moved to the front half of the aquarium. The overgrowth and roots mean that the discus can be seen more in the foreground, which is more attractive to the observer. In this way, it can be deliberately ensured that discus are always visible and that they get used to swimming in the foreground.

What needs to be considered?

The most important rule is that the discus fish that are introduced should not be too small. It is a mistake to introduce small discus measuring just a few centimetres into a display aquarium and then try to grow them up. This does not work. Small discus need to be fed often and thoroughly in order to grow. They need regular partial water changes so that the nitrate content of the water remains very low. Nitrate in the water is a growth inhibitor. This would mean that water would have to be added to the display aquarium far too often. This causes unrest, the plants grow poorly as the water parameters are constantly changing and other factors also play a role.

So buy larger discus! Buy discus at least twelve centimetres in size.

If you can go a little bigger, that's even better. One-year-old discus with a body size of 14 to 16 centimetres would be ideal. These are more problem-free, forgive feeding errors and will still gain a centimeter or two in size. In any case,

it is better to put only two or four large discus in such a display aquarium instead of having to deal with a dozen small or half-grown discus, some of which would certainly fall by the wayside.

Another important piece of advice is to only put suitable roots in the aquarium.

Of course, an aquarium with roots looks nicer. But the risk is too great to begin with if unsuitable roots cause rot in the aquarium. There are special stonewood or bogwood, but also very lifelike artificial roots made of safe material. Ask for them in the pet shop!

In any case, water new roots a few days in advance and keep renewing the water.

To give an aquarium background a modern look, the backlights from Daytime can be a completely new aquarium decoration for back walls. Naturally adjustable and a new dimension for the back wall.

The right location

An aquarium in a living room needs the right location. If there is too much natural light, there will immediately be algae problems. If the aquarium is next to a door, the fish can be frightened more easily. Lots of little things that need to be taken into account. Now you have decided on a furnished aquarium! You‘ve envied your friend‘s beautiful aquarium for long enough and you‘ve wanted one yourself for just as long. It‘s clear that it has to be an aquarium with discus fish.

Visit a specialist store to find out which suitable aquariums are available on the market. There are a variety of combinations with and without a base cabinet. Specialist manufacturers naturally also offer custom-made aquariums.

The size of an aquarium must be matched to the size and quantity of the discus fish. There is no question, for example, that six large discus fish cannot be housed in an aquarium of less than 300 liters.

If the aquarium is to be installed around a corner or somewhere else, it must also be ensured that there is enough space to carry out the necessary maintenance work without having to climb over obstacles.

The larger an aquarium is, the easier it is to maintain over the long term. For us discus enthusiasts, it is therefore better to plan generously here. The standard height for system aquariums in specialist shops is 60 cm. Bear in mind that the substrate and top edge of the water will reduce the water height by a few centimeters. You will find that a discus aquarium with a height of 70 or even 80 cm makes a great aquarium that is very easy to set up and can become a real show aquarium. Add to this a depth of 70 or 80 cm and your aquarium becomes a super aquarium. In terms of length, all options are open to you for custom-built aquariums. You can order such an aquarium from a specialist retailer according to your requirements, which is no longer a problem and delivery via a shipping company is very easy. Special sizes of any kind can be ordered from your specialist retailer via

Tropical-Deutschland, a specialist for such special aquariums. However, as the light intensity can no longer be sufficient from a height of 60 cm, Daytime LED light bars are the ideal lighting for such large aquariums, as lights up to four meters long are available to buy in the system.

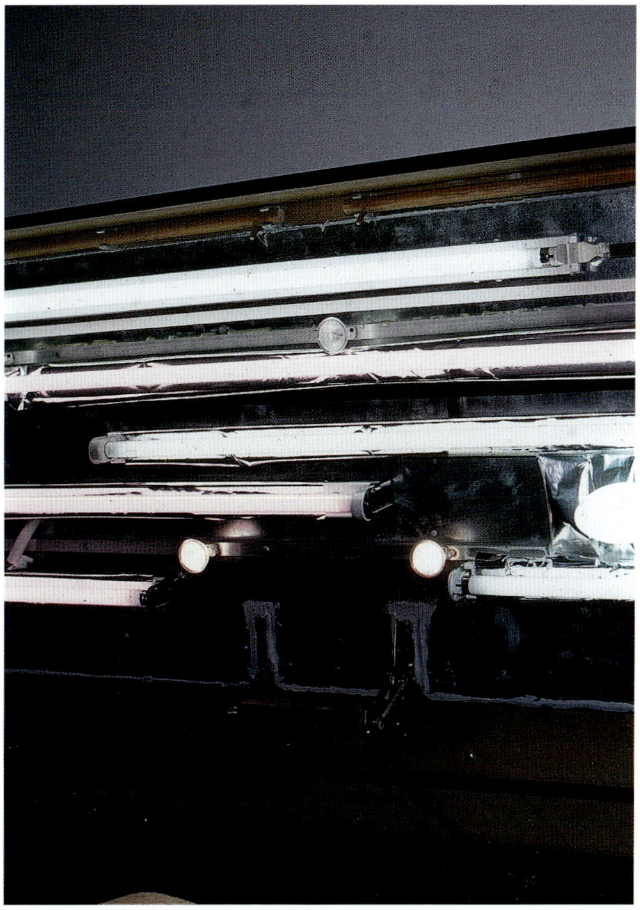

Neon tubes have had their day. Their time is up, both in terms of durability and light quality.

Suitable equipment

You have selected and set up your aquarium. Now you can slowly start setting it up. For an aquarium to look good, it must have a beautifully designed back wall. There are fantastic backgrounds in specialist shops that look so natural that they give the aquarium a wonderful ambience. If these back walls are too expensive for you, you can also install simpler colored back walls or, for example, cover the outside of the back wall with a colored plastic film. However, do not choose a color that is too dark, as dark back walls absorb a lot of light. A medium blue or a light green and a lighter brown have proven to be effective here. Plant pictures are also available in specialist shops, which can then be bought from a roll in the appropriate length and attached to the back wall with adhesive tape. A lot has happened in recent years when it comes to back walls. You should simply take a stroll through a specialist pet store and take a look at the corresponding decoration options.

As far as the right substrate is concerned, the discussion is perhaps already underway. Nature provides us with fine sand: Brazil‘s rivers often feature huge white sandbanks with the finest sand you can possibly imagine. Of course, the water conditions in such a river are very different from those in a small aquarium. However, river sand has the advantage that no food resi-

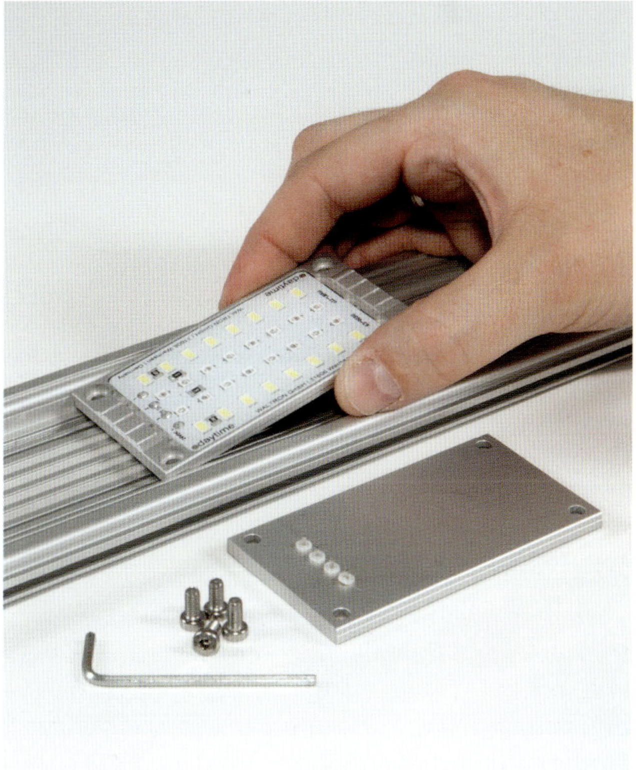

Modern light bars have individual LED inserts that can be fitted with a wide variety of LEDs. This makes it easy to install lighting effects. The entire Daytime system is fully compatible like a modular system and has all the advantages of modern lighting.

dues can accumulate in it, as is the case with coarse gravel. If you want to use gravel, it should also be as fine as possible, as discus fish do not like coarse substrate. They love to blow water into the substrate, stir it up and look for something to eat.

A good partner for such a fine substrate is underfloor heating. First, thin heating cables are laid evenly across the substrate and attached with rubber suction cups. The heating system glued to the glass bottom in this way ensures slow and easy water circulation in the substrate. This prevents rotting. In a figurative sense, a bottom heater works like a slow-working filter. Underfloor heating is particularly advantageous for planted aquariums.

If such a substrate heating system is installed, the height of the gravel or sand substrate may be six to ten centimeters. This bottom heating can be combined with an additional thermostat heating rod. A thermostat controller for the heating is recommended for large aquariums anyway.

How might a modern discus display aquarium be decorated? What does modern even mean? Modern would certainly be the realization of the natural conditions in discus biotopes in their native waters.

Large, even very large discus display aquariums, which are mainly decorated with bizarre roots, are always attractive. Muted colors are typical features and you should definitely avoid light-colored limestone in a discus aquarium. Dark granite, slate slabs and large river pebbles are perhaps the most beautiful types of stone to decorate a discus aquarium.

Modern aquariums are now fitted with LED units.

The popular Java fern

Which discus aquarist is not familiar with the Java fern, Microsorum pteropus? Java fern is so robust that this must be the main reason for its popularity, because if you were to succeed in ousting this attractive plant from your aquarium, you would have to make a lot of mistakes. It is therefore a plant that must be made for aquariums, because Java fern not only copes with a wide variety of light conditions and water qualities. Incidentally, hardly any fish or water snails will eat these fern fronds, which makes them even easier to care for.

The Java fern belongs to the spotted fern family and is widespread in tropical Asia. Due to its native habitat, it is therefore not a typical plant for our discus aquarium, but that doesn‘t really matter. After all, we are looking for hardy, resistant plants for our discus aquarium that have few requirements in terms of water quality and lighting. The leaves grow from a creeping rhizome and are slender and lancet-shaped, especially at the beginning. The roots of the Java fern cling to many furnishings and continue to grow from there. It is therefore easy to clamp roots between stone slabs or on decorative roots and simply wait until the roots of the fern have found a firm hold. Initially, you could support this, for example by tying it down with some fishing line or simply a rubber ring. The rubber ring and string can of course be removed once the Java fern is firmly established. Entire back walls can be covered decoratively with this fern. For example, a cork back wall covered with Java fern could make a great impression in a discus display aquarium. Java fern reproduces very easily by means of adventitious plants on the leaves. This means that small plants with their own roots form and grow like offshoots. Java fern can also be easily propagated by simply dividing the rhizome. If the plants are to be placed in the substrate, a loose substrate is recommended, whereby the roots should initially be weighted down with a few stones until they have found a good hold.

There is a large-leaved and a small-leaved growth form on the market. These ferns are also often tied to roots and sold as overgrown roots. This looks very attractive and allows numerous furnishing options. For example, the decorative bogwood root can be combined with the decorative Java fern.

For breeding aquariums in particular, which are usually very bare, it would be worth considering using a Java fern like this in combination with a bogwood root as a decorative item.

Perhaps your discus fish will then spawn on this bogwood root. Such a root is also popular as a shelter if it is attached to the top of the aquarium with a suction cup. The discus fish can now stand directly under this rootstock with the Java fern. This is very much in keeping with their natural behavior. Java fern can easily withstand temperatures of up to 30 °C and this plant can also cope with soft discus water.

The genus Echinodorus with its sword plants offers suitable aquarium plants for discus aquaria. However, they do require a little more attention in terms of care and aquarium design. Several species are found in the Amazon basin, but are now "produced" by the thousands in aquatic nurseries. Large-leaved sword plants are very suitable as individual plants in discus aquaria. Smaller species can be used for group planting on the substrate.

Which filtering should be preferred?

Filtering a display aquarium is of course necessary, but the filtration is very different from that of a heavily stocked rearing aquarium with discus fish or a holding aquarium that does not contain any furnishings. We actually want to copy nature in a planted display aquarium and in nature the water is constantly filtered very slowly. For example, rainwater seeps very slowly through the various layers of soil and is filtered in this way.

During this seepage process, oxygen is removed and the water is given a low pH value. On its way through the soil, the surface water is enriched with carbon dioxide and naturally dissolves minerals from the soil.
Bacteria convert mineral salts into plant nutrients, which the aquatic plants can later absorb

In this very large aquarium, a large group of Pigeon Blood Discus will look good. Java moss with roots, a large Amazon sword plant and some Vallisneria on the water surface are enough to create a perfect picture. The fine substrate is ideal for discus aquariums.

again. Nutrient-rich and healthy water is also required in the aquarium for good plant growth. This can certainly be achieved through proper filtration. In the mostly sterile discus breeding aquariums without plants, oxygen is supplied by the filter. Oxygen input through fast-running filters with a strong inflow of water is therefore very important here. If, on the other hand, an aquarium is planted and furnished, it is exactly the opposite, because then no additional oxygen should enter the aquarium. Slow-running filters with a weak outflow are the best solution here. So-called biofilters, which are placed under the aquarium, are often used. A wide variety of filter materials are placed in different filter chambers, which guarantee such slow and gentle filtering, whereby the pollutants are broken down by bacteria. Every filter should also remove dirt from the aquarium, which is why it often makes sense to combine mechanical high-speed filters with biological low-speed filters. This can therefore mean, for example, that a small rapid filter is used from time to time alongside a large, slow-running biofilter in order to visually clean the water. Regular partial water changes are very important to support the filter performance. Ideally, a large partial water change should be carried out once a week. This means that around 15 to 30 % of the aquarium water should be changed.

In nature, most aquatic plants thrive in slow-flowing or stagnant water. This is also where our discus fish feel most at home, as these aquatic plants and the roots of trees provide ideal shelters and breeding grounds for the discus fish. In the filter, it is not the existing filter substrate that cleans the water, but actually the bacteria that break down the pollutants in the water.

This is also one reason why complete filter cleaning should never be carried out. The entire filter substrate should never be replaced at once or even scalded with boiling water. Enough old filter material colonized with bacteria must always remain in the filter so that bacteria can quickly colonize the new filter material. For new set-ups, it is therefore ideal if it is possible to remove some filter material with bacteria from filter units that have already been run in. Such inoculated, new filters work much faster and the waiting time until the first fish can be introduced into the aquarium is also reduced.

Large colorful discus in a planted living room aquarium are a great sight. It is therefore always better to choose adult discus for such a display aquarium. A few red neon as a splash of color are a good addition. C. Nüss

Coarse soiling can be avoided by using a pre-filter made of filter floss. The filter floss catches the coarse contaminants and once a week this filter floss is simply thrown away. The fine, almost invisible dirt particles pass through this filter floss into the lower layers of the filter where they are converted by the bacteria. For a large display aquarium, slow-running biological filters with overflow protection can be installed under the aquarium. Especially if you place your aquarium in a cabinet, you have every opportunity to place such a biofilter in the base cabinet.

So-called multi-chamber internal filters can also be glued into very large aquariums.

This is often done in a side panel or in the corner of an aquarium. If the pump that pumps the water out of the filter is not too large, it can also be filtered slowly and gently. An addition to such a slow-running filter would be the so-called rapid or pot filter, which can be used externally at all times or as required. Do you want to make things very simple for your discus display aquarium? Then just use the existing tap water. Too bad if you live in an area where the water is very hard. Then there may be problems. However, there are no problems with medium-hard water. Up to a total hardness of the tap water of around 10 °dGH (degrees of German hardness) and a carbonate hardness of less than 8 °KH (carbonate hardness), the tap water is completely

Aquarium landscape with very intensive planting, modeled on the modern "aqua scape".

unproblematic for such a display aquarium. However, if the values rise significantly above this, it may be advisable to carry out partial demineralization. The simplest way to soften the water is certainly the reverse osmosis system, which is already available at very reasonable prices. A small reverse osmosis system is also completely sufficient for one or two display aquariums. Only if you intend to set up several discus aquariums and perhaps breed discus fish at a later date will you need to consider purchasing a larger reverse osmosis system. The reverse osmosis system is simply connected to the water pipe and works without any problems. The water is only divided into demineralized and salted water by the mains water pressure. The desalinated or almost desalinated water is then used for partial water changes in the aquarium. Blending with partial osmosis water is often sufficient to achieve the desired water values. The residual water that occurs as waste water can be used normally in the garden.

The purpose of reverse osmosis is to significantly reduce water hardness, which goes hand in hand with the removal of salt. Bacteria and pesticides are also retained, which is also a positive effect. Undesirable nitrate is also suppressed and reduced. This makes it possible to mix the right discus water, which both the fish and the plants need for optimum growth, using very simple means.

When choosing your first stock, it is good if you can select your own initial stock from a larger group. One such young discus per 50 to 60 liters of aquarium content is enough to start with.

Coarse gravel is completely unsuitable for discus aquariums.

Problems in everyday operation

There will always be minor or major problems from time to time. In most cases, it is the keeper's keeping errors that become apparent. New plants or new fish are simply bought and placed in the aquarium without checking or quarantine. If a display aquarium is working, nothing should be changed if possible. Many minor problems can be solved by a large partial water change. Please note: filter the fresh water through activated carbon once! This will remove harmful and irritating substances from the tap water. Nowadays we no longer know what is floating in our tap water. Suddenly a discus turns dark, settles, hides and no longer eats. This is usually the result of stress. A pair of discus has got together and is now suppressing the other inhabitants. This leads to stress. Suddenly one discus is suppressed and no longer feeds well. Now the intestinal parasites, which are somehow present in the digestive tract, have the opportunity to multiply and cause problems for the fish. Or the discus fish suddenly become shy, hide too much in the plants and perhaps only come forward to eat. This is where an intervention in the decoration of the aquarium can help. Thin out the plant population. Fish always tend to hide. So fewer plants, fewer hiding places. However, you must always check the water quality. Use water test strips, which are available from various companies.

This is a quick and easy way to check the water quality. Plan a night light. Even the tiniest night light is better than no light at all. There are very nice little LED lights. They are ideal for placing in an aquarium cabinet.

However, newly set up aquariums are often stocked far too quickly. The impatient owner wants to put in his newly purchased discus straight away. They simply do not wait long enough and the aquarium has not yet settled in. Initially, too much nitrite builds up and if the fish are then introduced too early, they will have serious problems. If the fish are then uncontrollably treated with medication for supposed diseases, the discus will really get into trouble unnecessarily. So wait and see, check the water values and only introduce the discus fish when everything is in order and the aquarium is working well.

The partial lighting is downright spooky.

Moonlight creates atmosphere! The modern Daytime LED lights have a perfect and professional control system to create such moonlight effects.

Any lighting mood can be created with modern LED lights.

In general, the control system makes it easy to set different lighting moods and to control plant growth in a targeted manner. With modern LED lights, aquarists should finally forget about neon tubes, also to save energy.

My show aquarium

Don't we aquarists all dream of a great display aquarium? In a way we do, because looking at magnificent discus fish, the kings of the Amazonia, together with some suitable peaceful fish friends in a beautifully furnished aquarium with plants and perfect light in the living room is extremely calming and satisfying.

Custom-made room divider including base cabinet from Tropical Deutschland GmbH with the following dimensions: 280 cm x 80 cm x 80 cm

My discus friend Marco Salafia has realized this dream perfectly and he shows here his show aquarium with exact instructions for a possible replica, even if it may turn out smaller or different. This room divider will definitely whet your appetite for aquariums. Aquariums as room dividers can be easily viewed from both sides. This is very interesting and also has the advantage that no aquarium rear wall is required. One disadvantage, however, is that turbidity in the water can be noticed very quickly, precisely because of the "see-through" effect!

Technology

Filter

Filter tank measuring 150 x 40 x 40 cm, fitted with filter floss, Hamburger mat, Syporax, boxtech filter blocks, activated carbon and zeolite. Return pump with 4500 liters return flow/h (calculated on the delivery height approx. 150 cm); located under the aquarium.

UV clarifier

90 watt UV clarifier from the pond supplies with a flow rate of approx. 1000 liters/h also calculated on the delivery height!
Important: no bypass, but direct flow.

Control heater

Two 300W Eheim heating elements (one in the filter shaft in the aquarium, one in the filter tank)

Lighting

LED lighting (Daytime LED Matrix). Lumen calculation, adjustment to aquarium size and installation by Giga-Zoo in Grafling. Fluorescent tubes were not considered, as this technology is now obsolete in aquariums. Daytime's LED units provide a completely different, perfect and adjustable light. The durability compared to tubes is remarkable.

Top-up water and water changes

- Change 30% of the water every 10 days and approx. 25 liters daily due to evaporation!

- Water change from the tap (Bavarian Forest with KH3 from the tap, therefore no treatment)
- Refill water only osmosis water from “Osmoseanlage 600 commercial” from Osmo Clean

Substitutions
- Sea almond tree extract prepared from leaves (approx. 200 ml per week)
- easy life plant fertilizer (approx. 30 ml per week)

Stocking:

Discus fish
- 6 Tefé Green Spotted (wild-caught) - Symphysodon aequifasciatus
- 8 Blue Moon Heckel Nhamundá (wild-caught)-Symphysodon Discus
- 5 Red Turquoise Red Spotted Symphysodon Aequifasciatus, captive bred

Corydoras
- 25 x Corydoras Sterbai - Orangefin Corydoras

Accompanying fish
- 150x Paracheirodon Axelrodi Red Neon
- Approx. 100 dwarf shrimps that feel at home in the filter tank.

Plants
- Limnophila sessiliflora
- Alternanthera reineckii
- Vallisneria gigantea
- Vallisneria nana

L-catfish
- 10x L 236 (Hypancistrus sp. - Super White Pleco)
- 10x L 201 (Hypancistrus sp. - Orinoco Angelfish Catfish)
- 10 x L 264 (Leporacanthicus joselimai - White-fin Cactus Catfish)
- 1 x L 14 (Scobiancistrus aureatus - Sunshine Pleco)
- 2x L 144 Ancistrus Sp. - Snow White Ancistrus
- 1x L 137 Panaque sp. Bruno - Blue-eyed Panaque = brown Bruno

Food:

Granules:
Self-made mixture of different Tropical granules:
- Discus Gran D50 Plus
- Discus Gran Wild
- D-Allio Plus
- Pro Defense
- Soft Line America

Flake food:
- Tropical D50 Discus Flakes
- Tropical Astacolor Flakes

Frozen food:
- Once a week, home-made beef heart mixture with spirulina and garlic
- Frozen, thawed Artemia crustaceans once a week

There is always movement in this show aquarium. Watching the young, lively discus grow is very satisfying. An aquarium with a clear view has a special charm.

Aquascaping

Aquascaping is the aesthetic realization and design of underwater landscapes in an aquarium. In the aquarium, landscape sections are recreated that are similar to those found in nature. The Japanese artist Takashi Amano was an important teacher of this technique. Alexander Paulus shows two examples of a discus aquarium here. Natural structures should be recreated using roots, wood and natural stone in combination with suitable plants. Ingredients from the real habitats of the fish do not necessarily have to be used. A depth effect in the aquarium is also desired.

Another wish for the show aquarium had to be fulfilled. Here, discus fish were to move into an aquarium 200 cm long, 60 cm deep and 70 cm high and feel at home. A touch of aquascaping was required, which was fulfilled perfectly. The aquarium was cleaned and then it was time to get started.

The first step was to add lava gravel to the glass base. Lava gravel is placed under the gravel in places where there is a build-up of roots or stones, so that the substrate remains active, even in places where there is a lot of build-up. Lava gravel has a very high colonization surface for good bacteria, which help to prevent rotting in the soil. The higher the structure, the higher the lava gravel can be used to gain height at the back of the aquarium. A substrate of fine gravel with a grain size of 1-2 mm is placed on top of the lava gravel.

The stone structure consists of Galapagos stone. Also known as "Galapagos Rocks", a porous, dark rock that actually comes from the rugged coasts of the Galapagos Islands. It is a natural product that is characterized by its unique, holey structure and is used for aquarium layouts full of character. Mining does not appear to be a problem. The stones do not release anything into the water and have no sharp edges, making them very suitable for a discus aquarium. Red bog roots form a very attractive color contrast.

Planting

- Lagenandra meeboldi - Mebold's Lagenandra
- Anubias barteri sp. Coin Leaf
- Bolbitis heudelotii - Congo Water Fern
- Cryptocoryne balansae
- Cryptocoryne beckettii Viridifolia
- Cryptocoryne wendtii green
- Microsorum minor
- Echinodorus reni
- Anubias congensis round

These are very easy plants to care for.

Stock

- 9 Discus
- 60 Neon

- L-catfish: L 200 HighFin Baryancistrus Demantoides - Green Phantom Pleco
- L 46 Zebra Catfish - Hypancistrus zebra
- L 264 White-Fin Cactus Catfish - Leporacanthicus joselimai
- 50 Steel helmet snails
- 7 Reticulated Siamese Algae Eater

Lighting

Daytime Matrix, maximum 10 hours/day

Filter technology

Filter tank (100/50/50) with 3 chambers
1. Mechanical filtration
2. Biomechanical
3. Biological chamber with Aquarock, a micro-porous filter rock

- Drain in the tank only with overflow comb Pump Aquabee 8000 / approx. 6000L / hour
- UVC lamp 36 watts connected via its own circuit

Fully automatic water change, 50 liters daily, controlled by a GHL computer. Fresh water connection directly in the base cabinet.

No CO_2 fertilization required in the aquarium. Fully automatic fertilization with fertilizer from C&P Aquaplants.

UVC clarifier

Discus community aquarium

Stones, for example Loa Rocks, are used with Yati wood as a base.

A community aquarium for discus fish with Altum angelfish and a few smaller fish, beautifully planted and decorated with a few roots is a dream of many aquarists. With an aquarium measuring 160 cm long, 70 cm high and 50 cm deep, a successful aquarium can be created. A so-called aquascape with sufficient retreat options for small perch and catfish, as well as for one or two discus. The dominant focal point was created by the overgrown rootstocks. A larger plant section was created in the right-hand section of the aquarium, which even grew into a small thicket on the far right.

This meant that all the inhabitants could be accommodated. When keeping discus fish together with angelfish, it is essential to ensure that the aquarium is large enough! A Fluval FX6 filter with biological and mechanical filter materials was used for the technology. Occasionally supported by an 18 watt UV-C clarifier. An Oase CrystalSkim 350 runs at night for better gas exchange and a slight water movement.

An aquarium computer regulates the pH value via CO2 to a value of 7.0. The aquarium is illuminated with Daytime Pendix LED pendant lights.

Slow-growing plants were preferred for planting, such as: Anubias nana, Pinto; Anubias nana, Microsorum pteropus, Bolbitis heudelotti, Helanthium tenellum, Crytocoryne crispulata var.

Aqua Substrate (or similar) with fine 1-2 mm natural gravel is used for the substrate.

The plants are placed in the gravel. In some cases, roots can also be fixed under stones or on the roots to allow them to grow.

View under a root. This will later provide an ideal hiding place for smaller fish.

Discus fish in a planted aquarium are always a particularly beautiful sight.

An imposing male discus on his quiet foray through the aquarium.

This female discus has developed a particularly strong red coloration.

Altum angelfish have become roommates with the discus here. There is no tension, the angelfish are already "old" acquaintances of these discus from another aquarium.

The two discus swim calmly and relaxed through their refuge.

Would you prefer a natural aquarium?

Endless expanses, huge river systems, tropical jungle - this is how we should imagine the Amazon. Quite different from a planted aquarium with discus fish. What does a natural aquarium for discus fish actually look like? This question is not so easy to answer, as we would first have to find out exactly where discus fish live in nature. Are they found on the banks of huge rivers or perhaps only in the small tributaries of large river systems? Do they perhaps live exclusively in the floodplains? However, this is not really possible, as there are also seasons when the river landscape is not flooded. Wouldn't they then have to live in the middle of the rivers during the dry season? Are there any aquatic plants in these rivers? Or are there perhaps mainly floating plants? Do discus fish really only live among roots or could it not also be the case that discus fish pre-

Sand, leaves and wood form the basis for a typical natural aquarium.

An artistic Back to Nature back wall with filter chamber is a good decoration for the natural aquarium.

A wonderful group of beautiful discus has lined up here for the photo.

fer to live among the leaves, branches and tree trunks of submerged, flooded trees and undergrowth? Lots of questions - hardly any answers. After all, who really went to the trouble of traveling to Amazonia, spending weeks there looking for discus fish and then finding out exactly where and how they actually spend their lives? Actually, it is more conjecture when such claims are made. Shouldn't aquarists be given the right to design their home aquarium with discus fish as they wish?

Does it harm the discus fish if they swim in an aquarium with gravel, even though they often live on river sand or a silt bottom in their native habitat? Does it harm the discus fish if giant vallisnerias form a wonderful back wall in the aquarium and these giant vallisnerias were not to be found at all in their home habitat or should it perhaps be a few Echninodorus Amazonas sword plants that have legitimately found their home here in the discus aquarium? Isn't it all a bit exaggerated?

If we were to recreate a natural aquarium as it actually occurs in nature in a river section of the discus biotopes, it would be a sad aquarium, because in nature there are often few aquatic plants or only one type of aquatic plant for miles on end. Unlike in our aquariums, ten or twenty different plant species can coexist.

Many years ago, I brought original Rio Negro sand from the home waters of the Heckel Discus and added a layer of sand about five centimetres high to my breeding aquariums. All the fish visibly felt at home on this light-colored sand and it could even be said that spawning behavior has improved significantly. This was also noticed by a large professional breeder who only offers his

Discus and angelfish can also be kept together under certain conditions.

discus pairs for sale in aquariums with sand bottoms. There they spawn as quickly as possible. It would be worth a try. Discus fish love fine sandy soil because they can search for food in this fine sandy soil, which is typical for them when they blow into the sand with their mouths to blow out small crustaceans, worms or other food animals. You can observe this behavior very well in your home aquarium. Sand, bogwood roots and one or two plant species are the perfect basis for a beautiful and natural discus aquarium.

When critics claim that discus fish should be kept in plant-free aquariums, the easiest way to answer them is to say that discus fish live in submerged trees in the wild. These submerged leaves, branches and trees are the same as a furnished and planted aquarium. So there is nothing wrong with a few plants in a natural discus aquarium. When setting up a discus aquarium, a distinction must also be made as to whether it is to be an aquarium for wild or captive-bred

Due to the high stocking level, this aquarium is not a natural aquarium.

Java fern tied to roots is fine, but coarse gravel is a mistake!

discus. Breeding discus have completely different requirements in terms of water quality, for example. While wild discus should preferably be kept in aquariums with very soft water and a relatively low pH value, this can usually be neglected for captive-bred fish. Aquariums for captive-bred discus can therefore be designed differently. This is because furnishings such as sand, gravel or stones often release hardening agents into the water. Incidentally, the water in furnished aquaria is much more stable than the water in sterile breeding tanks. To set up a beautiful natural discus aquarium, you must adhere to the golden ratio. The idea of the golden ratio was developed and adopted by architects, artists and philosophers and has been used and further developed time and again in architecture and art.

It is based on the realization that hardly anything in nature is completely symmetrical. So don't place the eye-catcher in your natural aquarium right in the middle of the aquarium. Such an eye-catcher could be a nice large root, for example.

Such roots should always be placed slightly away from the center of the aquarium. When setting up a discus aquarium, you should also take the liberty of using aquarium plants that are not native to Amazonia, as it is particularly

Altum angelfish can be socialized well in a large discus aquarium. Vallisnerias provide good hiding places.

Natural aquarium for green wild-caught fish.

Java fern always combines well with roots.

important to use hardy and slow-growing aquarium plants in discus aquariums.

Let's get away from the idea that discus fish always have to be kept in sterile aquaria to prevent diseases in the fish. Discus fish will probably be much healthier in a natural aquarium with its large biomass. Regular partial water changes and, of course, checking the water values are important in a natural aquarium. Finally, treat yourself to your furnished discus aquarium, even if it does not correspond to the actual living conditions in the Amazon.

My original RIO NEGRO aquarium with fine sand imported in a container from the Rio Negro. Three pairs of discus populated this aquarium with a capacity of approx. 2,800 liters and a 480 liter filter tank underneath. It was therefore the easiest aquarium to maintain that I have ever owned.
▼

The right discus food Optimal feeding

Even Heckel wild-caught animals can be fed by hand if they are used to the steady hand of their owner and, above all, if the food tastes good.

Feeding discus fish can sometimes be quite a difficult task if discus do not eat or only eat poorly. Feeding used to be a laborious task, as the food was usually prepared by the fish themselves. Today, everything has become much easier, as there is perfect industrial food that is on a par with pure live food in terms of quality and often has the great advantage of a better balance of nutrients.

How can you feed your discus fish optimally? Quite simply, with the right food. Whether you feed your discus fish predominantly with ready-made food from the factory or with frozen live food is a matter of opinion for the owner, as you can feed your discus optimally with both feed components. Only the overall balance must be ensured.

It would therefore be wrong to feed red mosquito larvae day after day or, conversely, only discus granules. Theoretically this is possible, but in practice this one-sided feeding would have many disadvantages. So it's the variety that counts. Polyunsaturated fatty acids are important for a good discus diet, as they are also constantly recommended to us humans by nutritionists.

These unsaturated fatty acids are easier to digest and hardly put any strain on the organs. Highly saturated fatty acids, such as those found in beef heart, lead to organ fatty degeneration, which can be fatal. However, feeding beef heart and other fish meat or crustacean additives is very popular in South East Asia and, as can be seen from the very large, bulky discus fish, is also very successful. Extremely large discus fish are selling points.

In addition to these essential, i.e. vital, unsaturated fatty acids, carotenoids play an important role in feeding. Carotenoids are a large group of yellow, orange and red pigments that are responsible for some of the vibrant colors in nature's color palette. Think, for example, of the red tomatoes, the beautiful orange carrots, apricots, peppers and also the delicious red lobsters that we enjoy. All types of crustaceans are also an important part of the diet for discus fish. This is why high-quality foods also contain crustacean meal and high-quality fish meal. On the one hand, carotenoids improve the color of the fish, but on the other hand they naturally also ensure better health and reproduction. This is usually forgotten when it comes to the color of fish. Carotenoids are so-called antioxidants, which significantly reduce the risk of disease when consumed in optimal doses. There are more than 600 different carotenoids in nature, some of which can be converted into vitamin A.

The best and most thoroughly studied is beta-carotene, which is also the best known. Discus fish naturally have a considerable amount of red color cells, but these are visible to varying degrees. Naturally, turquoise discus fish also have such red color cells - and depending on the type of food, these color pigments are more or less pronounced. If the food contains sufficient carotenoids, these red color pigments increase slowly but continuously. This is clearly visible and is noticeable in the bright red color of the fish. For example, turquoise-colored discus show strong red-colored lines or dots.

In green Tefé wild-caught fish, for example, the weakly colored red belly dots are much more prominent, making these fish considerably more interesting. The turquoise or blue markings remain clearly visible.

When feeding food with a high carotenoid content, it is possible to enhance the color of the discus fish in a natural way without the fish looking unnaturally colored, as would be the case if hormones were used.

When hormones are used, the fish become unnaturally reddish all over their bodies and this red coloration, including in the tail fins, is a clear sign of hormones. Such hormone discus should be rejected as they also lose the color again and then, in addition to the loss of color, there is also possible damage to the organs. Such fish treated with hormones tend to become infertile.

At discus shows, artificially coloured discus fish are quickly recognized by experienced judges and disqualified or devalued in such a way that they cannot take any of the top places. The use of hormones, such as testosterone, causes the color to intensify very quickly. This color enhancement occurs after just a few hours and the discus fish suddenly show much more color. The testosterone is then usually simply added to the aquarium water. Dealers used to claim that this would allow customers to see better what color their purchased discus might be later on. But in the end, it was just an attempt to trick the buyer into thinking they had more color so that they would be willing to pay the asking price. The young discus fish that showed such unnaturally strong color lost it again after a few days when they swam in another aquarium without hormones. Now the buyer rightly felt deceived and the trust was gone, even if the discus showed more color again in the course of growth and grew into well-colored discus.

Today, hormone supplements in the food are no longer an issue, as the color foods in particular are so well balanced that the natural colors of our discus fish can be seen well even without such negative additives. However, the aquarist must also be careful not to feed exclusively colored foods, as otherwise an excessive feeding of colored foods would also lead to an exaggerated coloration of the discus fish. The use of foods containing carotenoids or food mixtures with added astaxanthin should always be viewed positively, as the supply of carotenoids improves health and also spawning behavior. It should also be noted that carotenoids improve egg formation in female discus and increase courtship motivation in males. With natural carotenoids you are doing something good for your discus fish and you should definitely include this in their diet. Carotenoids are not always red, which is why foods with added algae, such as spirulina, have become an integral part of a modern feeding plan. Spirulina algae in particular has become the food of choice for discus lovers. Modern industrial feeds can offer a great range of balanced and finely tuned foods that leave nothing to be desired.

Feeding from the hand

With training and patience, you can also get your discus fish to accept food directly from your hand. Simply take some of their favorite food between your fingers and hold it in the water. The fish are curious and will come swimming towards you. Of course, there is no practical benefit to feeding from your hand. But it is great fun and gives satisfaction in this beautiful hobby. Almost all types of food are suitable. For example, cubes of frozen food are very easy to take between your fingers and offer. You should also try large flakes of food. Even wild-caught fish can't say no after a period of acclimatization and will eat a lot.

Food envy in the group plays a role that should not be underestimated.

Larger groups of discus will gather around the keeper's hand more quickly and greedily accept the food. Individual discus are certainly more cautious. Feeding from the calm hand of the keeper is also a good way of getting unusual types of food "to the fish". Give it a try, it's fun.

But don't give up disappointed after the first few attempts.

Healthy discus fish are not shy and will soon eat greedily from your hand.

Industrial feed

In terms of quantity, industrial food is the most consumed food in the aquarium hobby. Industrially produced foods are all the many types of food that are offered in cans, bags or other containers for the convenience of the aquarist. Feeding convenience is one thing, optimum nutrient supply is another.

The variety of standard and special foods on offer is huge.

Initially, discus larvae feed on the skin secretions of their parents after swimming free. But after just a few days the little discus are already looking for the finest food in the aquarium. Freshly hatched Artemia crustaceans are the first food. The new Insect Menu in size XXS is a perfect industrial food that is precisely tailored to the needs of these small discus. It is best to feed them in addition to the Artemia.

Insect Menu as finest granules XXS

There is a whole range of special discus foods for discus fish.

Flakes are certainly ideal and are especially popular with smaller discus.

Reddish foods are usually preferred, but discus also like to eat green spirulina flakes. These should definitely be included in the diet as premium food. Granulated food is also standard for discus. They should only ever be fed as much as they can consume in a few minutes.

It is better to feed more often and in smaller portions. Food tablets stuck to the glass are an ideal food for smaller discus. A few drops of vitamins can also be dripped onto the tablet beforehand, which makes the whole thing even more nutritious.

Freeze-dried mosquito larvae or crustaceans make sense, but are often not eaten readily. You can now safely raise your discus on a variety of ready-made foods and feed and care for them perfectly throughout their lives.

Live food

The pros and cons of live food will always be discussed.
Proponents swear by the hunting instinct of the discus, which would be preserved. On the other hand, a discus will certainly chase after a flake of food floating downwards. Live red mosquito larvae are always available in pet shops.

The larvae, which are popular with fish, can be fed after being well rinsed.

Young discus in shoals eat flake food or granules very greedily. To begin with, also feed food with added algae and not just beef heart mixtures.

They are also recommended in the literature as medicine carriers. Live midge larvae are then bathed in a water medication solution and are supposed to absorb some of the active ingredient, which the discus then eat. A little laborious, but it can work. Live, home-bred enchytrae are a good food variety and their nutrient content can be easily manipulated by the food you put together.

For the discus damselfish, freshly hatched Artemia crustaceans are still the food of choice anyway. However, the shells must be sieved off when feeding.

Breeding the crustaceans is very easy and you can buy breeding starters in the shops.

Frozen food

The range of frozen food is also huge. Every pet shop has a freezer in the store containing numerous types of frozen food. The most common type of frozen food is probably the red mosquito larva. Eaten by all types of fish, it has become a top seller. Unfortunately, the quality of red mosquito larvae varies greatly and it can only be recommended to pay attention to quality. Poor quality red mosquito larvae can quickly cause problems in the discus aquarium. It doesn‘t have to be the case, but it is possible that red midge larvae are contaminated. As they particularly like to live in dirty, smelly water, they are not actually the ideal fish food. White and black mosquito larvae are less harmful, but are not eaten as readily.

Enchytraeids are a very good live food, but they are a bit of work. You need to buy a breeding stock and care for and feed it. You can then remove these small worms and offer them as a well-accepted food. I like to use industrial flake food as food for the enchytrae, which enhances the quality of the enchytrae.

Freshwater cyclops is a very small food that the smallest young fish can soon eat. All crustacean-type frozen foods are recommended, but are accepted differently. The discus must first be trained to eat them.

Krill has a somewhat harder shell and therefore discus like to refuse this good food.

It is always pointed out that frozen food should be fed thawed.

Home-made food

Making your own food is an old tradition. Today, however, it is no longer worthwhile for most aquarists, as there is enough pot food on the market.

For discus enthusiasts, home-made food mash is still an issue when breeding. Although it is no longer necessary here, the old tradition is still followed.

Homemade food is actually always frozen food. Beef heart is often the basis of the mash. This is never entirely uncontroversial, as meat from slaughtered animals is not considered suitable as food for fish by scientists. The digestive tract of the discus is not actually adapted to beef heart, but can still cope with it. Feeding beef heart alone, without any ingredients, would not be okay. The fish would lose their color and would certainly develop deficiency symptoms. This means that other ingredients have to be added to a home-made feed mash.

Fish meat and shrimps, for example, would be ideal. Fish meat from lean fish and prawns can be used both raw and cooked. Whether you peel the prawns or use them with their shells also depends on the size of the prawns and how finely they are ground. Spinach is often added and there is plenty to choose from in the freezer. Some manufacturers add eggs or just egg yolk, others add a little dissolved gelatine to thicken. Others swear by paprika powder as a coloring agent and so on. The addition of mineral powder and a multivitamin solution certainly makes sense.

The mash is finely mixed, frozen in plastic bags or shallow containers and fed as required. Good preparation of the beef heart is important in any case.

The fat must be trimmed off. The large tendons should also be removed.

The chilled, diced heart is then minced, mixed with other ingredients and frozen.

Live enchytraea are a very popular food for discus of various sizes. However, these small worms should not be fed every day, but should remain a special meal for the discus. Also a good food for a breeding pair, especially for female discus, to encourage spawning.

Successful discus breeding

The pinnacle of discus care is the successful breeding of discus fish. Fifty years ago, it was sensational when an aquarist succeeded in breeding discus fish and raising the small discus. This is a difficult undertaking, because although we have now uncovered all the secrets of discus breeding, it is still a great challenge to raise the discus fry to the point where they can be passed on to other aquarists. However, this success can be topped off with a wild-caught pair that could be persuaded to breed and then several dozen small discus swimming freely in the aquarium. Do you want to give it a try? It's worth it, because it's very fascinating. Wild-caught discus often have even more intensive breeding care than captive-bred discus.

Once a pair of discus has finally been put together and transferred to the breeding aquarium, they will soon start preparing to spawn. The two fish court each other and swim towards each other. In doing so, they bow to each other and sometimes the males also poke the females lightly in the side. These small scuffles are completely harmless. It is only advisable to separate the two partners if there are serious disputes that last for days. It often helps to separate the aquarium with the help of a wire or glass partition so that the fish can see each other but cannot fight. If it is a pair, the fish will slowly get used to each other.

As a rule, courtship is peaceful and there is no danger to either fish. The appearance of the discus fish will also begin to change. As they prepare to spawn, they change color. In technical terms, the darkening of the discus fish in the rear half of the body is called "sooting". This typical spawning coloration is intensified by a stronger protrusion of the last four to five vertical stripes. These signs indicate that spawning is imminent. The discus pair is looking for a suitable place to lay their eggs. When breeding discus, clay spawning pots that resemble inverted

Breeding is a fine art, as it requires a sure instinct to raise half a dozen or more fry to breeding maturity. This has been achieved well with this group of Virgine Red Scorpion.

A breeding picture from the old days, when it was almost sensational when such successes were possible. Even today, the breeding and above all the rearing of small discus is proof of perfect aquaristics.

The next step is to put together a breeding pair. Ideally, a pair should find each other and perhaps start preparing for spawning in the community aquarium.

burial vases are preferred. These pointed clay vases are an ideal spawning substrate as they are somewhat rough and the eggs adhere very well. However, egg laying is not dependent on the presence of this clay pot. Other objects are also suitable as spawning substrates. It is possible to use upturned clay flower pots, sloping roof tiles, plastic plates, plastic pipes, bricks or other materials. The only important thing is that the material used does not release any harmful substances into the water. If these objects are missing, the discus fish will spawn directly on the glass of the aquarium or on the heater if necessary. The selected spawning substrate is now cleaned vigorously by the two fish. They use their mouths to thoroughly clean the place where the eggs are laid. This cleaning takes several hours. It is accompanied by vigorous shaking of the body. The cleaning and shaking are typical signs that a spawning run is imminent. Discus fish usually spawn in the evening. This spawning in the evening is particularly true for discus in breeding aquariums with a water

This pair is already beginning to darken in color, which is a good sign that they are preparing to spawn. A clay vase in the aquarium is usually used as a spawning site.

The pair stands at the spawning site and begins to clean it. The discus shake again and again, almost excitedly. This behavior is a lot of fun to watch.

temperature of around 30 °C. As the water temperature has a very strong influence on the development of the eggs, they need just under 60 hours to develop in water that is 30 °C warm. This means that the eggs, which were laid in the evening hours, are so far developed in the morning hours of the third day that the larvae hatch out. Now the discus parents have daylight available to continue caring for the larvae. The free swimming of the discus larvae also takes place during the day, making it easier for the parents to collect the wriggling fry. Many discus breeders offer their breeding pairs a night light. This means that a small lamp burns near or above the aquarium throughout the night. This gives the discus fish the opportunity to look after their clutch and fry better. Specialty stores sell special energy-saving lamps for this purpose.

So-called moonlight in the form of blue fluorescent tubes is also suitable as a night light. Abruptly switching the light off or on causes all fish to become frightened. This could have fatal con-

In the case of discus that are valuable to the breeder, small clutches are also obtained in order to obtain offspring. Normal clutches contain at least 100 to 200 eggs.

Discus are often not so choosy when it comes to choosing a place to lay their eggs. Professional Asian breeders also offer glass plates that can be easily removed from the aquarium for artificial breeding or nurse breeding, for example.

This clutch is defective. The white eggs are moldy and others are just empty egg shells. There are several reasons for this, for example unfavorable water conditions, unfertilized eggs or even planaria that eat the eggs.

sequences, especially for discus fish with eggs or young fry. In extreme cases, the disturbed fish could eat the eggs or fry.

Finally, spawning is imminent - now the female is the more active part. The male usually stands slightly away from the spawning substrate and observes the female and the surroundings. It is therefore also important that visual contact with other aquaria is interrupted. It would be unfavorable if the male discus were distracted by rivals in a neighboring aquarium. After all, it should fertilize the laid eggs immediately. As the egg membrane changes after just a few minutes and becomes impenetrable to the sperm, this is really important. The female becomes increasingly restless and begins to swim to the spawning site again and again to test spawning. No eggs are laid yet. This test spawning can also take place over a longer period of time. Finally, the female swims from the bottom to the top of the spawning site and begins to lay the first string of eggs.

The eggs have already developed here, as the dark spots in the eggs indicate larval growth. A filter pipe was simply assumed to be the spawning site here.

The discus parents become darker due to the nutrient mucus and the larvae eat this skin secretion as their first food. Well-maintained pairs separate and the larvae constantly alternate between the parents.

Ten to 20 eggs spill out of the laying tube and adhere firmly to the spawning pad. The female swims away from the spawning pad and allows the male to fertilize the eggs. Ideally, the male should be waiting next to the spawning female ready for fertilization. The male's sperm is only viable in the aquarium water for about 30 seconds. It is therefore very important that the male now releases the sperm over the eggs so that the sperm filaments can penetrate the microphyle of the egg. Incidentally, females always lay their eggs with the microphyll facing upwards. Logically, there should be no strong current in the aquarium during the fertilization process, as this could impair the flow of sperm. So if you can observe the mating, you should interrupt the filter circulation during egg release and fertilization.

After fertilization of the eggs, the male swims a semicircle and returns to the female, which is already laying the next egg string. Depending on the age and condition of the fish, such a spawning phase lasts about an hour. During this time, the female lays 150 to 400 eggs, with an average clutch containing around 150 to 200 eggs. As discus fish can spawn again at intervals of five to seven days, the size of the clutch depends heavily on the physical condition of the female.

If the discus fish eat their eggs again after spawning, they will produce a new clutch after just five to seven days. This can be repeated so often that an enthusiast may despair. If the larvae do not hatch, the discus fish will lay eggs ten to fifteen times in succession. If the pair raises their larvae and the fry then remain with their parents for three to four weeks, they will generally only spawn again after the fry have been caught. However, it can happen that the female carries larvae that are a week old and still spawns again. If this happens, the fry are usually eaten by the parents. It is better to remove the new clutch, as the parents may then take care of the existing fry.

The clutch is taken from a second pair and placed with a pair that is taking good care of it. Both broods are then raised together. This nurse care also ensures that the pair without a clutch spawns again quickly.

It happens again and again that the parents eat the eggs. The reasons are not known, only assumptions. Once the larvae swim freely, rearing works. This is why the clutches are protected by grids. In breeding farms, the clutches are usually immediately protected by grids.

Part of the clutch is defective and has turned white, but nevertheless many larvae have developed and hatched. To count the larvae or eggs more accurately, simply take a photo and count them at your leisure.

The clutch is largely fungalized. It is advisable to check the water parameters. Osmotic pressure, i.e. water that is too hard, destroys clutches. The conductivity and pH value of the water are important measurements.

Immediately after spawning, the aquarium filter should be switched on again. After the eggs have been laid, the two discus fish take it in turns to stand in front of their clutch and fan the eggs with their pectoral fins. This fanning is intended to prevent the eggs from turning fungus. The fanning provides the clutch with more oxygen. If there is a night light above the aquarium, the fish are encouraged to constantly and devotedly look after their clutch of eggs. Egg eating is a phenomenon that has not yet been unraveled.
This egg-eating usually only occurs on the second day after the eggs are laid. It can therefore be assumed that the eggs did not develop properly. For this reason, they were then eaten. If fertilized discus eggs cannot develop, this is usually due to unfavourable water conditions. Discus fish are probably naturally equipped to recognize when their fry have a good chance of surviving and when they do not.

Towards the end of the second day after egg laying, dark nuclei can be clearly seen in the eggs, indicating a fertilized egg. These eggs have developed normally and already after 55 hours a clear movement can be seen in the eggs. When viewed with a magnifying glass, you can even see the beating of the heart and the tail of the discus larva. The larvae begin to hatch with the tail first. They break through the egg shell. It takes another 60 hours for them to develop into swimming larvae. Now the adults transfer their larvae. This means that they suck the larvae - which beat wildly with their tails - out of the egg shells and deposit them in another place in the aquarium. The larvae have an adhesive thread on their head, with which they adhere very well to the spawning substrate. Repositioning and sucking out is also a disinfecting process. The larvae cling to the spawning substrate with the adhesive thread until their development is complete. Then these adhesive glands on the head atrophy and the larvae can swim free. The parents tirelessly pick up the free-swimming larvae with their mouths, chew them up a little and then spit them back into the crowd of wriggling larvae. Anyone observing discus fish looking after their larvae for the first time will become very uneasy and think that the larvae are now over. However, the parents do not eat their young, but chew through them for several seconds and then spit them back onto the spawning substrate. This is therefore a normal part of the brood care of these cichlids. During the time of this intensive brood care, the parents only need to be fed once a day. Even when feeding, one of the parents remains with the fry and guards them. A well-harmonizing pair of discus will therefore take turns both caring for and feeding the fry.

Six days after the eggs are laid, the larvae are able to swim. On this sixth day, their motivation to swim away from the spawning substrate begins to increase. Now so many larvae swim free that the parents can no longer prevent it.

Rearing the fry

It worked - the discus larvae have swum free. The most important thing now is for them to find the parents and swim towards them to eat their first food. As we know, the parents produce a slimy skin secretion during this time, which is urgently needed by the discus larvae as their first food. This skin secretion contains important bacteria, but also enzymes, which the small discus fish need as a starting aid for their digestive system. As the formation of this skin secretion normally results in the parents looking much darker, it is important that there are no other dark objects such as filter sponges in the aqua-

This harmonious pair leads a group of young fish through the aquarium on both sides. The small discus can stay with the parents and continue to eat skin secretions until they become a nuisance to the parents or they want to spawn again.

A very harmonious situation. The young actively eat supplementary feed, because the light modern colors produce significantly less skin secretion and therefore supplementary feed is required very early on.

A group of Yellow Pigeons that have developed very evenly, which is a good sign. When buying, look for signs such as growth, eye size and body shape.

rium at this time. Otherwise it could happen that the fry swim towards these objects due to their motivation to feed.

Especially very brightly colored discus parents - for example Pigeon Blood discus, yellow or gold-colored discus fish sometimes produce less skin secretion. In such cases, the larvae may have problems swimming to the surface. As a result, some discus fry have starved to death in the aquarium within a few hours.

Normally, the larvae swim to the parents and start feeding immediately. This feeding can be observed very well, as the small bellies of the young discus quickly fill up. They graze their parents almost tirelessly. The two parents also take turns feeding the young fish. Pairs that harmonize well develop a handover system that works very well. Through twitching and jerky movements, the parents communicate commands to their young or hold them close to their bodies. If one of the parents wants to hand over to its partner, it begins to twitch more strongly, then suddenly swims away and literally shakes the fry off its body. These now swim quickly to the second parent nearby and continue to feed there without any problems. During this phase, it is very easy to feed the parents, as they know how to distinguish food from fry - it is not uncommon for them to eat their young. The keeper will certainly not feed the parents too often during this rearing phase. One or at most two smaller feedings a day are sufficient at this stage.

The small discus fish can be fed with freshly hatched Artemia crustaceans just five days after swimming free. The aquarist normally leaves the young with the parents so that they can continue to eat the secretion. Secretion and newly hatched Artemia crustaceans are now the main food. After hatching in salt water, the small crustaceans must be rinsed well with fresh water so that the young fish do not ingest too much salt.

The small discus are very quick to chase after and eat the live crustaceans. To make feeding these crustaceans easier, it is a good idea to suck them into a pipette and then blow them directly into the school of young fish. It is also possible to suck up the Artemia crustaceans with an air tube and then feed them directly into the school of fry.

During this time, the water level can be lowered slightly, which makes feeding much easier. This trick is also used if the fry have difficulty swimming to their parents. If the water in the aquarium is drained to such an extent that the parents can just about swim upright in the aquarium, there will logically be less water in the aquarium and the young will have less swimming space, making it much more likely that they will have to swim up to their parents. Some breeders also darken the aquarium considerably in the first few days after free-swimming in order to attract the young fish to the parents' bodies through the movement stimuli they receive from them. Once the small discus fish have eaten plenty of Artemia crustaceans, almost nothing can go wrong with their further breeding.

Some breeders swear by separating the young fish from their parents after just seven or eight days and rearing them with Artemia crustaceans and Moina (puffer flea). Of course, this is a little more work, but it also ensures that the transmission of diseases from the parents to the fry is reduced. If you now want to separate the small discus fish, which are not even one centimeter in size, from their parents, you must use appropriately small aquariums so that you have all the fry together and they are properly fed. Feed at least six times a day to ensure that the little ones grow optimally. However, this frequent feeding also ensures that the water quality constantly deteriorates and therefore daily partial water changes are actually mandatory. During these intensive partial water changes, the food residues are sucked out at the same time.

On the other hand, you can also leave your small discus damselfish with their parents. This allows the parents and fry to spend another four to five weeks together. It is only advisable to separate parents and fry when the small discus fish eat the parents' skin so aggressively that injuries occur.

After four to six weeks at the latest, the fry will have reached a size of around one euro coin and will now require a corresponding amount of space. The larger the rearing aquarium can be, the better the fry will grow. Of course, the food quality must also be right. Regular partial water changes are an important tool for accelerating growth.

So-called Ica x Japura Heckel are sought after. The fifth black middle band in particular is a sign of these Ica-Discus. These are mixed crosses in which genuine Heckel ancestors were involved.

Heckel cross with the name Oriximiná, derived from the town of Oriximiná on the Rio Trombetas in the state of Pará in Brazil. Such names of rivers or fishing areas are often adopted in the names of wild-caught fish.

What is it?

A rare historical photo! I can't remember exactly when it was, but it was definitely many years ago, because these were the first real red spotted offspring of Kitti, which were also a typical Kitti sensation!

A pair of discus have found each other in the community aquarium and have proved to be very harmonious. It is then possible to relocate the two discus to a breeding aquarium. An aquarium with a capacity of 120 liters or more is sufficient. An air-operated sponge filter, as in the background, is ideal due to the mild current.

The typical clay burial vase was accepted as a spawning site. However, the majority of the eggs have already turned fungus. The entire clutch will probably be lost. Reasons could be poor fertilization, but also water problems. Low water conductivity values are important for discus breeding. A conductivity value below 200 microsiemens gives the best chance of breeding success, as this reduces the osmotic pressure on the egg shell.

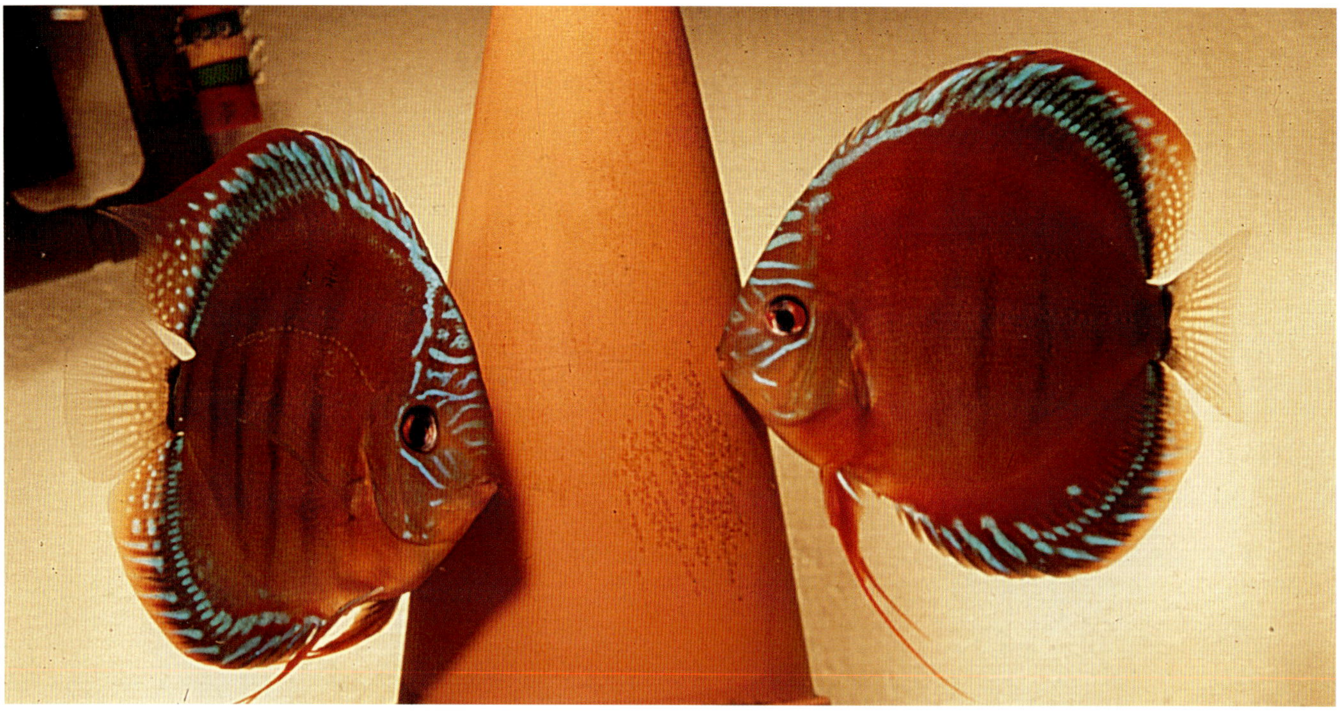

Both discus show the usual darker vertical stripes. The clutch is very fresh and the eggs are clear. Egg numbers in discus fluctuate in the normal range between 100 and 250 eggs, but can also reach up to 400 eggs per clutch. Factors such as food, well-being and age of the females certainly play a major role. As the brood develops, the parents become darker and darker. This dark coloration is a color signal for the hatching fry to better find the parents with the skin secretion and to be able to swim to them safely.

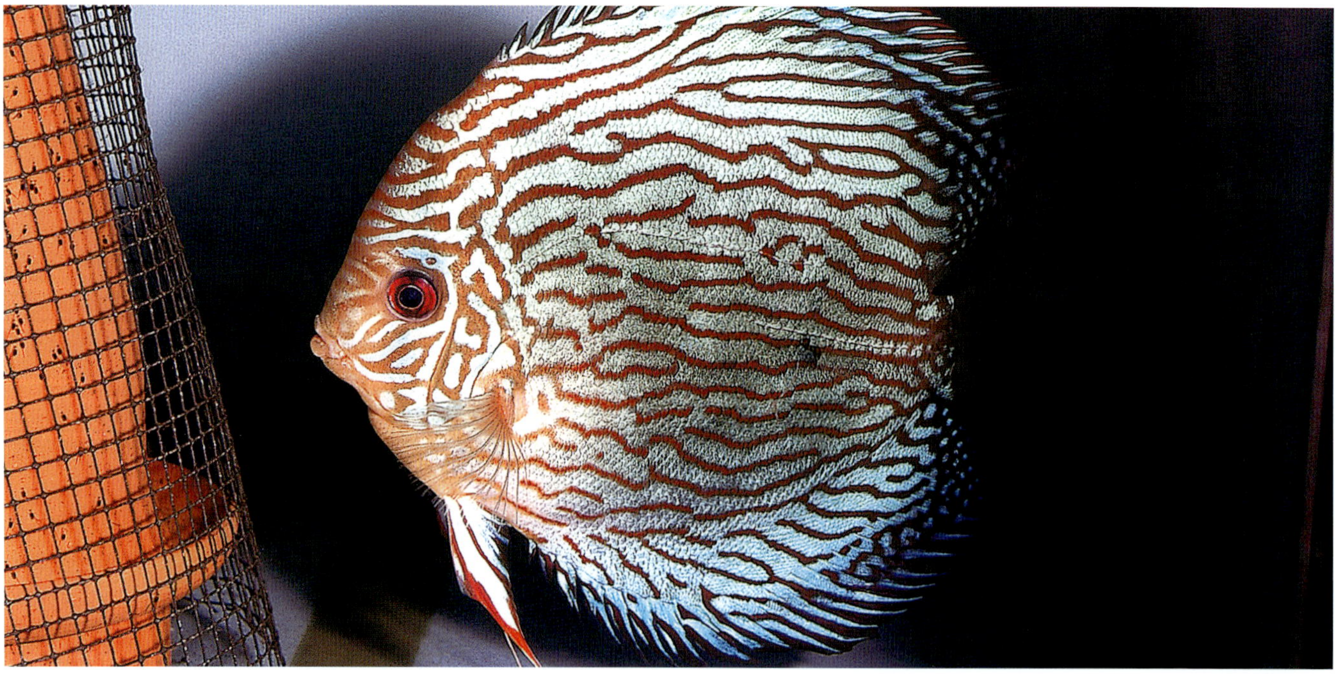

It happens that discus eat their clutches again and again. The real reason why breeding never works is not known. Suspicions point to inexperience on the part of the fish or incorrect water values, so that the discus already suspect that egg development cannot work. Assumptions but no proof. This is how egg protection with wire has become established. This can be seen very often in Asian breeding farms. Usually as a preventative measure, regardless of whether the discus are egg eaters or not.

Here, a pair in a larger aquarium chose a filter tube as a possible spawning site. The male was actually significantly larger, but this does not always have to be the case. Body size is never a clear sexual characteristic, especially in captive-bred discus. In discus captures, the males usually develop more color and body fullness in a larger group.

The wire remains on the pot until the larvae have all swum free and the parents have swum on. The wire and pot can then be removed from the aquarium. The skin secretion is not so clearly visible in cobalt blue turquoise, but it is present. The slight dark coloration is already sufficient for the larvae to find the parents under normal circumstances. If there are problems, it would be a good idea to attach a small spot of night light under which the parents stand so that the larvae can find this spot of light on the parents.

White discus, a rarer breeding form, usually have small clutches and also produce less skin secretion. Perhaps a precaution taken by Mother Nature. Skin secretion is still present, but less. It is a good idea to lower the water level so that the fry are closer together. The swimming space is narrowed. Provide the fry with their first fine food such as Artemia nauplii as early as possible. The new Tropical XXS Insect Menu would be ideal as a second food, as it already contains 45 % real insects.

The young nestle close to the body. There are certainly more fry on the other side of the body. The partner keeps a close eye on what is going on with the school of young fish. If one discus wants to release its young, it begins to shake its body and quickly swims towards the other discus, which now takes over the shoal. A well-rehearsed pair will lead their young for four weeks without any problems. However, if they want to spawn again earlier, they start to clean and the young must be caught.

Suddenly such so-called Ica wild-caught types appeared, which showed the black Heckel stripe. The fifth stripe is not actually more visible in blue, green or brown wild-caught specimens. However, if it suddenly appears in such wild-caught specimens, they can be mated with other discus. These extreme peculiarities ensured that captive-bred discus of all varieties were mated with these Ica types. This resulted in more and more color forms that showed these striking black stripes. The result was an absolute bestseller that is still very much in demand today. At the same time, new trade names also emerged. Today, the naming of discus fish has become very confusing.

A well-maintained pair takes turns peacefully guiding and feeding the young fish. The golden discus shows very little skin secretion, while the other discus is very dark and full of skin secretion. The juveniles shuttle back and forth between the parents and also react to their signs. For example, the parents call by twitching the young close to their bodies. The older the young fish get, the further they move away from their parents and search for food in the aquarium on their own.

This pair does not even seem to have a dozen young, but this is still very important for the breeder because it is a sought-after mating. The golden Pigeon Blood and the almost white partner can again help to create a color variant that can become very popular in breeding. This is a principle of breeders to always offer their customers new color nuances. In this way, new trade names can quickly be added to the price list and the business with new color variants can continue. This is how modern discus breeding works today.

A cobalt blue flat turquoise that caused a stir in the aquarium scene 30 years ago. Today it is a simple standard discus that would no longer meet higher demands. The first thing to mention is the missing red eye. The body shape is not perfect, the forehead too steep. The fact that the blue color appears a little too dark could be due to the fact that this discus is ready to spawn or does not feel very comfortable in this aquarium. The coarse gravel is atypical for discus aquariums. A fine sandy bottom is always preferred. Time passes and the requirements change considerably. Forty years ago, I saw the most magnificent large cobalt blue discus in Hong Kong and was thrilled.

These golden breeding forms are very popular because they no longer show dark stripes. All imaginable color forms are now available in the trade. Their body size and shape are also impressive. The small kink in the right anal fin can be forgiven. Here too, as is so often the case with modern offspring, Pigeon Blood genes are involved. These discus are very hardy and easy to care for.

Visually a completely different caliber is this circular and very powerful red and white discus. This is not an albino. The white head section is a breeding result of the so-called Marlboro Discus from Kitti in Bangkok. The head is white and the rear half of the body is red. Just like Kitti's favorite cigarette brand. People like to show these discus with an even stronger red coloration.

Good half-grown discus that come from a breeding. They have different colors. On the one hand, the red spots of one parent were passed on, as well as red and white body colors. Desired colour characteristics could be strengthened and stabilized by backcrossing these discus. For example, inbreeding or line breeding with spotted or almost completely red pairs. A lot can be achieved through these gimmicks in large breeding facilities and the market can always be kept going with color variants.

Note the rings or circles in the upper edge. These were the first ringed Leopard Discus from Malaysia to appear. Later, more such closed rings were added to the pattern. This is how the name Ringleopard came about. The Leopard Discus differ from the Snake Skin Discus mainly by the number of nine dark vertical stripes they have. Snake Skins usually start with 13 stripes and sometimes even end with 16 vertical stripes. Depending on how you count the sometimes somewhat branched stripes, you could even count up to 17 stripes. It remains to be seen whether this still looks nice. It is certainly strange.

Nhamundá Red is a magic word for wildlife enthusiasts. Nhamundá is a small town on an island in the lake district of Lake Amazonas. And the area is known for very beautiful and interesting discus fish. Especially red-brown and really red-bodied discus come from here. After just a few days in the export stations, these discus quickly become stronger in their red tones if kept well and later, with optimum care in the home aquarium, they are very impressive beauties. In addition to the multi-surface red discus, there are also dramatically beautiful striped forms.

Here are two Snake Skin discs of a brown variant. Here you can see 15 vertical stripes. The discus are healthy and swim in a beautiful, large aquarium. Color is a matter of taste and this uniform brown gives the viewer little color. But they do look special with these many stripes. Due to this brown coloration, the individual scales do not show the dramatic pearl pattern.

Nahmundá Royal from the area of the same name as the Nhamundá Red but with the typical lines of the royal varieties. These discus are usually referred to as royal when their entire body is covered with horizontal blue or turquoise lines alternating with brown-red or red lines. One could say that these discus are completely marked. Royal varieties are highly sought after and correspondingly expensive.

An attractive Royal Discus pair that is an F1 offspring. (F1 = first offspring generation) The female is actually somewhat slimmer and also has a round dorsal fin. However, these are not always unmistakable sexual characteristics. As F1s they still have typical wild-caught characteristics. In the wild, male discus are always somewhat more pronounced in terms of both color and other body features. Discus live together as a group of several specimens in river and lake retreats. There, the males are the more dominant specimens.

Can you think of a name for these discus? They could be called Red Sunrise or San Merah or even Golden Rose! There are many names and paper is patient. I met a so-called doctor from Asia on the Internet who lists 36 different discus colors on his homepage alone and offers the information for sale. That's great. I'm still an old-school discus enthusiast and, after all, with my magazines, yearbooks and discus books, which have been translated into 15 languages, I have disseminated more "discus literature" than any other author. I must and have had to accept the fact that so many fantasy names have emerged on the discus market for commercial reasons. Always buy the discus that you like and that makes you happy.

The discus in the future

Something completely new and chic. A strong silver colored body with beautiful red fin coloration. Whether it will be possible to stabilize this exact distribution of red tones permanently is questionable. A discus that will find its fans who are prepared to pay a high price.

Let's take a look back together at the past 60 years of the discus hobby, but also at the future that lies ahead of us.

When I took my first steps into the discus hobby, I had already spent several childhood years with aquariums and the then standard fish of guppies, platies and swordtails. It was fascinating to watch these livebearers constantly reproducing in the aquarium. The breeding almost went by itself and sometimes I dreamed that I could earn a great pocket money from the many offspring.

At some point, discus fish turned up at my pet shop and the asking prices were very different to the guppy prices. Perhaps this was the key to earning money with ornamental fish breeding after all? But what about breeding? Difficult! There was no specialist literature on successful discus breeding. Jack Wattley's offspring were unobtainable, Dr. Schmidt-Focke's discus were unthinkable. But there were a few good specialist stores that already sold discus offspring. When I moved to the Bavarian Forest and into a house with its own drinking water source in the garden, I came into possession of almost genuine Amazon water. Bavarian forest water with pH 5.6 and around 80 microsiemens. That was my chance to breed discus. That's how it all began and the first small discus book soon followed from my pen. Then book number two and soon, due to the great demand, number three! The third book was published by the major American publisher TFH with Dr. Herbert Axelrod and quickly became a worldwide success, as there was hardly any literature on the discus and its successful breeding.

bede Verlag became a specialist publisher for aquariums, then also terrariums and all other pets. In total, bede published over 600 pet titles. German discus offspring, especially the German Rottürkis became known worldwide and suddenly a discus hype went around the world. Particularly in South East Asia, many large farms were established that also bred discus fish. Cross-breeding successes occurred time and again and new color variants gradually emerged, which became established and are still popular today.

The Heckel Wildfang is currently very popular with crossbred discus because of its black fifth stripe. Since Heckels in the Amazon region have also mated with other color forms of the discus, such crossbred discus have come onto the market. Now the cross-breeding of the Heckel band is beginning in all possible offspring variants. There will be many new developments in the future.

As the sale of discus fish quickly became professionalized, new and perhaps even more striking discus had to be offered to the buyer again and again. This was and still is the reason why new color variations and new trade names appear. The classic and safe color designation for the purchase is no longer reliable. However, we now have the Internet, which means that up-to-date pictures of discus for sale can be sent around the world in a matter of seconds. This is a good way to order discus fish online from a trustworthy seller. Shipping options have also become much more reliable these days.

So it‘s best to have a look at your dream discus online beforehand and then have fun with this hobby. You can see some particularly eye-catching discus here and perhaps this is what the future of our discus really looks like.

Another discus of the future? A so-called Koi Discus. As far as the markings are concerned, you can accept the name, because there is a similarity. Through constant cross-breeding, mostly in larger Asian discus farms, there seems to be no limit to the imagination.

In the past and also in the future of discus breeding, there can always be strangely marked discus if random discus continue to be mated with each other. Let's take a look at a smaller Asian "discus farm" where production is the main focus. It was similar many years ago when the first Pigeon Blood Discus were bred in Thailand and it later turned out to be a genuine mutation.

Here is such a cross between Pigeon Blood and Cobalt Discus. ▼

▲ *A strange Snake Skin type appears here, which has a rust-red base coloration. It shows 14 vertical stripes, which is always an unmistakable sign with Snake Skin. Snake Skin are actually bred more in the classic turquoise or blue-red shades. Quite a striking discus.*

This blue Haraldi wild-caught fish has already received this Heckel stripe through cross-breeding and such wild-caught fish are often offered with the addition Ica Discus.

Crosses of offspring with wild-caught specimens. Here the Heckel stripe is clearly visible again. Trade names Ica x Japurá Heckel. Rio Ica and Rio Japurá are the names of rivers in the Amazon region.

Two further magnificently colored offspring, which are often offered as Koi Discus because of their markings. The imagination has found almost no limit in the names of modern discus colors.

Leopard Heckel are offspring of Leopard Discus that have been created by crossing Heckel Discus or offspring with Heckel stripes. They have retained the very strong coloration of the beautiful Leopard Discus, but also show this black band.

A rare discus with a dark blue fin edge attracts attention. Strong red and delicately distributed white scales enhance the color effect and make it a particularly attractive eye-catcher in the aquarium.

Years ago, there were silver Ghost-Discus in Penang - ghostly and unusually ugly. Now, similar variants with more color are appearing: such as the blue Ghost with a golden head or the Golden Ghost. They look mysterious and eye-catching.

Also a Ghost Discus, but with a silvery blue background. During a mating of a beautiful brilliant blue discus, a silver Ghost came in between and this was the result, much to the delight of the breeder. So a new Silver Ghost.

How should this discus be successfully named so that it becomes a bestseller? Perhaps Marble Discus, i.e. Marble, because of the interrupted and shifted black stripes? These are certainly accidental breedings and of course not mutations, as happened with the Pigeon Blood in Thailand.

Another chance find that I want to show here. You have to know that there are discus breeding farms in Indonesia and China that sell many thousands of discus every month. With such quantities, there are always special hits.

Blue Scorpion is the trade name of this attractive discus, which has no vertical stripes but still has the scales of a Snake Skin. Very good round body shape with good finning. Many discus of this quality are exported to us from South East Asia today.

Platinum Discus is the trade name, because platinum is a precious metal that ranks alongside gold at the top of the list of designations for particularly beautiful and valuable things. Then Platinum Discus must also be allowed, because these two are really flawless white-silvery shining, have no disturbing stripes, but strong red eyes.

The Discus love story

» Get to know

» Spawning

» Hatching

A special kind of love story, photographed by master photographer Horst Linke. Enjoy the process of discus breeding from finding the couple, their willingness to start a discus family together, caring for the clutch and larvae, protecting and feeding the fry together to raising the young fish.
Pure fascination!

» First steps

» In growth

» The offspring

The time has finally come for the female to lay her eggs on the cleaned spawning substrate. The male fertilizes this clutch alternately with the female and, apart from a few eggs, all seem to be well fertili-zed.

» Get to know

» Spawning

» Hatching

» First steps

» In growth

» The offspring

» Get to know

» Spawning

» Hatching

Both discus fan and guard the clutch. Some unfertilized or destroyed eggs turn white. Others develop well and a black nucleus becomes visible.

» First steps

» In growth

» The offspring

Depending on the temperature, the larvae hatch from the eggs after 50 to 60 hours. They remain attached to the spawning substrate with an adhesive thread on their head for a further two days. The parents usually reposition the larvae in the aquarium.

» Get to know

» Spawning

» Hatching

» First steps

» In growth

» The offspring

» Get to know

» Spawning

» Hatching

The free swimming of the fry is crucial for successful breeding. Now the little discus have to find their parents with the important nutrient secretion in order to start feeding straight away. These are the first steps or fin strokes.

» First steps

» In growth

» The offspring

» Get to know

» Spawning

» Hatching

You can clearly see that the little discus larvae have already filled their bellies with skin secretions. In the first few days they are very closely attached to their parents and do not leave their side. The larvae only become more active when they are fed with Artemia crustaceans and are already chasing them vigorously.

» First steps

» In growth

» The offspring

» Get to know

» Spawning

» Hatching

Thanks to the skin secretion and plenty of additional food, the little discus grow quickly. After just 10 to 14 days, they begin to grow a little taller.

Perfectly groomed pairs of discus do not quarrel and take turns in guiding the fry very well. They use jerky movements and twitches to communicate the appropriate commands to their school of young fish.

» Get to know

» Spawning

» Hatching

The ingredients of the skin secretion have been researched and it is important to favor this natural breeding. Artificial rearing is possible and perhaps even necessary in some cases when it comes to very important, problematic breeding experiments. Nothing can replace the fascination of natural breeding.

» First steps

» In growth

» The offspring

» Get to know

» Spawning

» Hatching